WINE
AND
CLIMATE CHANGE

L. J. JOHNSON-BELL

BURFORD BOOKS

For Michael, Charles, and Benjamin

And to the Winemakers

Printed in the United States of America.

10 9 8 7 6 5 4 3 2 1

Library of Congress Cataloging-in-Publication Data
Johnson-Bell, Linda.
   Wine and climate change / L.J. Johnson-Bell.
       pages cm
   Includes index.
   ISBN 978-1-58080-174-4
   1. Viticulture.   2. Wine and wine making.   3. Grapes—Climatic factors.
4. Terroir.   I. Title.

   SB388.J63 2014
   634.8—dc23

                                        2014015313

OTHER BOOKS BY THIS AUTHOR:

The Wine Collector's Handbook (Lyons & Burford)
Pairing Wine and Food (Burford Books)
Great Wine Tours of the World (contributor) (New Holland)

Foreign Editions:
The Home Cellar Guide (UK)
Good Food Fine Wine (UK)
Quel Vin pour Quel Plat? (France)
De Juiste Wijn bij het Juiste Gerecht? (Belgium)

CONTENTS

AUTHOR'S NOTE

Although I have titled this book *Wine and Climate Change*, its emphasis will not be purely on the topic of climate change, nor the causes and politics surrounding it. This is not a political work. It does not need to be. It is no longer debated that our planet is undergoing dramatic changes. The earth has always undergone weather cycles; she is a mass of ever-changing, ever-renewing energy. Both plant and animal species have experienced mass migrations due to climatic events since the beginning of the earth's formation. These cyclical events are simply becoming more intense and more erratic. The ensuing debate as to what impact human activities have made on these events, and to what extent mitigation is possible, is outside the scope both of this work and, quite obviously, I assure you, of my expertise.

This work is being told from a wine taster's perspective as opposed to a scientist's. It assumes that the reader is conversant in the relevance of *terroir*, the effect of temperature on grape harvests, and the distinction between good wine and fine wine. Most important, it will assume that the reader is interested not in proving climate change and its causes, and in divining the future, but rather in what is changing now, how wine producers are dealing with the changes, and what these changes might mean to the consumer. My emphasis will be on what the world's winemakers are trying to tell us. For they know better than anyone: They are on the front lines. To those few climate change deniers remaining, I would say that their ignorance is an insult to the millions of highly competent wine industry professionals who are up to their elbows in hard evidence. Climatologists love wine grapes: They are the most sensitive fruit crop.

As Dr. Richard Smart likes to say, they are the "canary in the coal mine." And what other farming industry has elevated harvest reports and tasting notes to such an art? This historical fodder alone provides the best possible research material.

Yet for some, it still seems to be difficult to connect the dots between extreme harvest conditions and climate change. I was invited onto *Sky News* in October 2013 to discuss the day's headline: the "global wine shortage."

All the copy coming out of the newsrooms was that "consumption was up" and "production was down due to bad harvests." Yet, nobody wanted to talk about the reasons behind the production decline. They only wanted to hear about how the Chinese are soon to be outdrinking us all: a far sexier headline. This was not the real issue. But the fact that wine production is down due to extreme harvest conditions that are only going to get more extreme was too scary for prime-time family viewing. So let's put a "happy ending" on the story. Yes, some wine regions, especially the fine wine regions of Europe, will change style. They are all gradually succumbing to irrigation, and the replanting revolution has already begun. Yes, some large swaths of warmer-climate regions will cease viticulture altogether after sustainability becomes unviable and their salinated soils can no longer support life. But we will see current cooler climate borderline regions improve in quality, in southern England and northern Europe. And some newer regions will emerge in northern Europe and in Tasmania and New Zealand as both hemispheres move poleward. For example, China is now our fifth largest producer; there are more plantings there than in any other country. So to try to blame them for drinking the world into a wine shortage is discounting the offset of their production. We will always have enough good wine, and we may be able to look forward to different fine wines. All that I hope is that the winemakers, who have the most to lose, are given the time, support, and assistance they need to adapt.

My twenty years of following wine harvests provide all of the firsthand evidence I need to convince me, and to illustrate the changes happening in the world's vineyards. You can taste climate change. Our wine regions are being remapped. That's not new; they have always been changing. But the changes were slower, permitting adaptation. Today we are witnessing an unparalleled rate of change. And never in the history of our planet has the wine industry been so firmly and comprehensively entrenched in our economic and cultural identity. These changes will have varying effects on different wine regions that will depend on their ability to change or adapt.

The thought of this both depresses and excites me. I already mourn the great Burgundies that first seduced me into this trade. And I am no longer able to enjoy my beloved Bordeaux—not at 15 plus percent alcohol. But I am enjoying tasting Pinot Nero from Alto Adige and the reds from Austria, and I am told that I should look forward to Swedish Pinot Noir. Let's see what happens . . .

Climate-change-related events and their effects on the world in general and on wine, specifically, are occurring every minute of every day. So I hope that this book will provide the background information needed to understand these changes and place them into a wine lover's context. I have incorporated a Wine and Climate Change Portal section on my existing website, www. TheWineLady.com, to contain videos, wine producer interviews, data, studies, and more . . . a sort of "live" bibliography, if you will. The scope of this topic is never-ending: There is so much more I could have touched upon, and there are areas that merit an entire book unto themselves. I am aware of this, and will endeavor to explore and expand upon them in future work.

Please visit www.thewinelady.com. And also feel free to contribute to the corresponding Facebook page: www.facebook.com/thewineandclimate changeportal.

L. J. Johnson-Bell
linda@thewinelady.com

INTRODUCTION

Iam standing under the shade of an olive tree that still clings to its sweetly pungent, oozing fruit. Sweaty grape pickers frantically race up and down the sloping hill before me. I am in the scorching heat of Emilia-Romagna, Italy's famed "fertile crescent"—its food basket. But the land beneath the vines is compacted and dry. Dust devils swirl between the gnarled roots. I am in the way and feel helpless. There is much shouting, and a busy hum with an undertone of worry. The owner looks tense and agitated.

The grapes need harvesting. *Subito*. But he has a dilemma. If they cannot get the grapes in fast enough, by hand, before the vines shut down from the heat, or the grapes burn, they will have to machine-harvest. He is against this. It bruises the grapes, and in this heat they would oxidize in a very short time. Plus, with all the bits the machine collects in addition to the grapes, triage back in the winery would be that much more labor-intensive. Still, better to machine-harvest than to let them die on the vine. But his planting density in many of his older parcels won't allow the space for a machine harvester—and even if he wanted one, there aren't any more available, as his neighbors are all experiencing the same panic.

His calculations tell him that his yields are already down by 20 percent, and it looks as though it is going to get worse. The heat means that his red varieties have matured at the same time, as opposed to in a staggered fashion, and he doesn't have the manpower to hand-harvest all the parcels. They can't seem to get the picked grapes into the winery fast enough before they are affected by the heat. He knows that many of his larger neighbors have invested in huge refrigerated trucks they park at the end of the rows so that the harvested grapes stay cool. He does not have this luxury. There are all sorts of practices he has had to consider in the past few years of drought and heat. Last year he had to resort to lightly irrigating some of the more vulnerable parcels for the first time. His list of compromises is growing, and it does not sit easily with him.

I can see him weighing the pros and cons in his head, his struggle with the gamble he is being forced to take . . . the added expenses . . . the possibility of changing his wine's identity forever. The taste of this wine that he has known since a boy hangs in the balance. I know that I am witnessing the involuntary abandonment of centuries of traditional and quality winemaking methods.

There are winemakers who hold on to "traditional" winemaking methods for too long and for the wrong reasons. Then there are the sort of winemakers who buy every new machine and adopt every new method, devoting themselves to maximum output with maximum marketing, dressing their wineries up as chemistry labs, forgetting that they are farmers, really.

But this winemaker is one of those who sit in the middle, wisely judging where it is logical to adapt and how to judicially apply new or different methods. A winemaker who puts the wine quality first. This temperate, Mediterranean climate has afforded him this liberty. Now, with the increasing heat, every year he is facing new problems and being forced to adopt methods that have already been embraced by the hotter New World wine regions from their inception, as their climates had not given them the choice. Now he fears that his wine, a wine that has been bred to speak of its unique *terroir* and to express its varietal character, will be suppressed, and forced to join the mass voice of the ubiquitous, international choir.

THE SECRET IS OUT

As recently as three years ago, if I were on a press trip and I or one of my colleagues brought up the topic of climate change, our questions were ignored and glossed over. But during my most recent vineyard visits, it has been the winemakers who bring the topic up. This could be simply because they no longer have any choice. The evidence is so physically visible. We are walking among shriveled vines and parched soils. The screaming headlines after the 2012 harvests alerted us to the fact that Europe is experiencing its worst grape harvest in fifty years. But for those of us who have been judging wines and visiting vineyards for twenty years (and more), this is not news. NASA reports that the year 2012 was the ninth warmest in their analysis of global temperatures that stretches back to 1880. In itself, that sounds fairly unremarkable, they remark. But as climate scientists note, what's important is the long-term trend. The 10 hottest years in the 132-year record have all occurred since 1998, and 9 of the 10 have occurred since 2002. "What matters is, this decade is warmer than the last decade, and that decade was warmer than the decade before," says Gavin Schmidt, a climatologist at NASA's Goddard Institute for Space Studies. "The planet is warming." This is manifested in the fact that, for example, Bordeaux's alcohol content has been creeping higher and higher for years now, due both to a desire to emulate the high-alcoholic (heat-induced) Napa wines so beloved by the American wine critics, and to having fallen victim to Mother Nature's unwittingly ironic plan to do it for them.

The New World wine regions of California, Australia, New Zealand, and South America have already been experiencing problems for much longer. These countries do not have indigenous grape varieties. The *Vitis vinifera* species was brought to them via the Europeans. Purists are perfectly entitled to argue that trying to grow grapes in a non-indigenous climate and soil was always going to end in tears. Australia is losing vineyards to extreme drought and rain conditions and has been producing hot, heavy, over-extracted brews for decades. Even allowing for natural variability, when paired with climate change, climate records get broken (Karl Braganza, *A Land of More Extreme Droughts and Flooding Rains?*, 2012).

But now, as the Cabernet-colored heart of our fine wine regions in Europe is finally hit, the issue has become mainstream and is no longer the preserve of the eccentric or scholarly few. The world is paying attention. It was acknowledged in the French press as early (or as late!) as 2003 that an unprecedented summer heat wave devastated the European wine production, which hit a ten-year low in crop yields. France suffered a loss of billions of euros. While the warming of the climate of Bordeaux, and other "then" cool-climate regions, in the second half of the twentieth century was welcomed for allowing more consistently ripe harvests and maturation, now the changes have swung the pendulum too far in the opposite direction and the effects are anything but favorable.

Winemakers have warned that the increasing number of hot days during *floraison* (the fruit's flowering season) speeds grape ripening, but not necessarily its maturation. These are two different things that I will explain later. This means a longer growing season and earlier harvest, which correlates with lower yields and poorer-quality grapes. Usually, low yields are considered a sign of wine quality; keeping yields down is a practice quality winemakers employ. Low yields are a good thing when they are a product of perfect climate conditions and expert vineyard practices. But when yields are rendered low due to extreme heat, drought, disease, rain, or hail, the fruit can be distressed or over-concentrated. This translates into unbalanced wines.

THERE IS SUCH A THING AS TOO MUCH VARIETY

Variable weather is good, extremes are bad. There is clearly a lot of emphasis on hotter vintages, but, in truth, what seems to be happening is that there are more extreme cycles of weather within the larger cycle of an overall warming. Harvest variations used to be the guarantee of wines with character and personality, but too much climatic variation means too much unpredictability and ruined crops. As NASA's Bill Patzert asks: "What is the amount of risk we can tolerate?" Harvest variations are the hallmark of Europe's fine wine regions. The challenge to work with or overcome Mother Nature is the "point" of the entire viticultural exercise. A great winemaker is one who can navigate the vagaries. The New World wine regions were pooh-poohed for being lazy: Wine growing in a constantly sunny climate is considered easy work—too easy—and the wines reflect that.

But now all is changing; too much is too much. Confirms Gregory Jones, "While 2010 was the warmest year on record for the northern hemisphere, more worrisome is the increasing climate variability—record cold winters followed by record hot summers, droughts and fire season giving way to extreme rainfall and flooding." Indeed, wine producers are experiencing extremes not only in one country—say, a north–south divide—but even in one region, and during one growing season. For example, extreme hail and snow in the spring, which destroys half the crop. Then, just as they think that they have recovered from that, along comes an extreme heat wave or drought before and during the harvest period that reduces what was left of the crop.

The United States has always had extreme weather, continues Patzert. "We look back on our weather history. It's been punishing: floods, droughts, tornadoes, hurricanes, great forest fires . . . Is global warming happening? No doubt about it. We're living in a warmer world, we're living in a melting world, sea levels are rising. We're seeing more frequent, more intense, and longer lasting heat waves. As far as hurricanes, tornadoes, forest fires, floods, and drought, the evidence is definitely not in."

At first, the previously much cooler and wetter climate regions, where it has always been difficult to mature white grapes, much less red ones, will flourish and enjoy a relative period of stability. But then as the hotter peaks create shorter, hotter maturation periods, raising sugar levels and lowering acidity, they will decline into non-quality. In Napa and Sonoma Valleys, the climate has become so warm that ripening fruit is not an issue. However, retaining acidity and developing flavor (flavor from the fruit and the soil, not from the selected yeasts used in fermentation, from over-extraction, or from new oak barriques) has become more and more difficult.

WHAT DOES THIS MEAN FOR US?

Why should we care about climate change and wine? Are you satisfied to just drink what will be offered up in the stores and not ask anything more? Climate change and wine is relevant to us all. Wine is made from grapes and grapes are fruit. In the same way that our food crops will be affected by provenance, quality, price, and scarcity (boohoo!), so then will our wine supply. We will have erratic wine supplies. Production costs for wine producers will skyrocket and so, too, will prices. Our favorite wines will taste different. I already have friends who ask me why they used to be able to drink an entire bottle of Rioja but now fall asleep after one glass.

Consumers will want to understand what is happening so that they can find wines they enjoy and understand why it is that they do or do not like a particular wine style. Some of our regions will change wine grapes and wine styles, or cease wine production. How we drink, buy, store, and invest in wine will all change. We need knowledge.

A case in point: A recent ad campaign by the Languedoc-Roussillon region stated that its wines are the "future classics" of France. These wines are increasing in quality, the campaign continued, and will soon be producing fine wine to rival France's other classic regions. Perhaps they are counting on Bordeaux and Burgundy losing a large portion of their market share. But what the campaign neglects to point out to the consumer is that in 2006, because of extreme drought and heat conditions, French president Nicolas Sarkozy was pressured by the region's wine producers to allow irrigation for the first time in French history. This practice is against French AOC and EU regulations—and indeed all fine wine production practices (more on this later). But once you have to irrigate, the fine wine game is over. Roussillon had no choice. They said that if they could not irrigate, there would be no crop of which to speak. Fair enough.

So bad are things that in December 2009, a group of over fifty French chefs, sommeliers, and château owners urged the French president to save the region by securing a deal at the international climate change conference in Copenhagen that year. They are worried that their wines are "increasingly marked by higher alcohol levels, over-sunned aromatic ranges and denser textures" and are losing their "unique soul." Heat does not make elegant or refined wine. Their petition asked for a global deal to cut industrialized countries' greenhouse gas emissions by 40 percent by 2020 and to set up "solid aid mechanisms for developing countries." They stated that if global temperatures rise by more than 2 percent before the end of the century, "Our soil will not survive and wine will travel 1000 kilometers beyond its traditional and indigenous limits."

Jean-Pierre Chaban, a climatologist at France's National Institute for Scientific Research, says, "We will have new wine-producing regions in zones where one doesn't normally cultivate vineyards, like in Brittany and Normandy and it will spread to Great Britain. One can even imagine vineyards in southern Sweden and Scotland." Julia-Trustram Eve of the English

Wine Producers adds, "Some say that by 2080, it will be too hot to grow grapes in southern England, where today, there are already 2732 acres under cultivation, an increase of 45% in the past four years."

Before I get too far down the "indigenous is best" argument, I should point out that "indigenous" becomes a rather comparative term when placed in the time line of the earth's climate. As the *Vitis vinifera* migrates and searches for pastures new, literally, it will want to move into habitats currently occupied by some of our endangered species and we will have difficult choices to make. It has done so before and will do so again. Says Dr. Rebecca Shaw of the Environmental Defense Fund's Land, Water and Wildlife Program, "Climate change will set up competition for land between agricultural and wildlife—wine grapes are but one example. This could have disastrous results for wildlife. Fortunately, there are pro-active solutions. We are creating incentive-based programs with private landowners to provide wildlife habitat as we expand our capacity to feed a growing planet in the future under a changing climate."

As adaptable as we are, things may move along too quickly to handle. Don't forget, growing grapes is not like the fashion industry. Trends cannot be accommodated for from season to season. Grapevines have life spans that reach into their fifties, sixties, and seventies, all the while producing better and better fruit. It takes at least three years before a vine is able to produce fruit and then many more to age and produce the best fruit. Quality wine producers in Europe often do not allow their vines to produce fruit until they are seven years old, and others still cellar and age their reserve wines for years before releasing them. Wine producers are going to have a very difficult time navigating their way through such a quickly changing landscape with a product that has a built-in time line dictated by nature.

Those in the hospitality industry will be keen to know where to be looking for the future quality wines and how to keep up with the changes so to better stock their cellars. They need to know which regions may no longer require cellaring, or may need less cellaring than before. They need to know which regions should not get away with asking high prices due to deteriorating quality and which ones should be supported.

Investors, particularly, need to know where to buy in the future and how to revalue their current stock. As the classic wine-investment regions continue to produce wines with 15 percent alcohol and higher, these wines will not have the aging potential they once had. So will this mean that investors will be able to invest less and recoup returns earlier as the wines will mature earlier, or will they lose their value entirely as they lose their once unique identity? Should investors start following cooler climate regions and buy wines that may turn out to be our future classics? Should they try to buy as much of our much older vintages as possible, knowing that this is their swan song? There is a lot to consider here.

And the most important consideration of all: the livelihoods of those involved in all aspects of wine production and the multibillion-dollar international industry upon which they rely.

YOU CAN TASTE CLIMATE CHANGE

The future? I am not a climatologist. And even the most expert among them is loath to predict the future. What is known is that there will be pauses and sharp increases and decreases. Mark Lynas states in his *Six Degrees* that "extreme weather has always been with us, but the fact that rising levels of greenhouse gases trap the sun's heat means that more energy is available in the system, so the worst is happening more and more often." Droughts will be hotter and longer and rain will be longer and heavier. So again, where "vintage variation" used to be what every quality wine producer sought, it is now the bane. The climate has become a quantity so unknown that it makes any reliable quantification impossible. These changes have brought more consistent vintages to some grape-growing regions, whereas in others they have brought great challenges: conditions that have been both too warm and dry, or more extreme or variable. Again, this book will place climate change in a wine-producing context. For each region, the answers will be different: a cessation of viticulture; an adaptation to climate change; an amelioration due to climate change. Some regions will newly emerge.

You can taste the difference the hotter temperature can make in a wine, just as you can distinguish between cool climate and warm climate wines. I am a completely objective observer. I have no agenda. I would like nothing more than to be able to turn the climate clock back and freeze-frame wine production somewhere in the mid-1980s. And I am sure that my older, more experienced colleagues may argue for an even earlier decade! Barring that, I wish nothing more than the preservation of fine wine production—to safeguard the wines that are distinctive, full of personality and a sense of place. I am not looking forward to a world increasingly awash with the homogeneous, bland, and international wines favored by the champions of mass-market commercialism. Like every other industry, wine is falling victim to sameness, whether due to an intentional stylistic choice, or one imposed by climate. We will always have enough good wine. Let's get excited about this pivotal era of the wine industry, about the potential for discovering new great wines. We have no idea what delicious new wines may be awaiting us. Among all those gathering storm clouds, there will be a silver lining.

What Should Happen in the Vineyards

WHY ADAM AND EVE REALLY
LEFT THE GARDEN OF EDEN

Int.   Terrace of a Sumerian Villa, Mesopotamia, circa 6400 BC—night

A warm, late-summer evening. The last rays of the sun illuminate the surrounding lush garden, laden with fruit trees and flowers. Mount Ararat looms overhead, a fertile Green Giant. A couple are preparing for a dinner party. She is setting out bowls of fresh figs, almonds, olives . . . He is carrying amphorae to the table. Both are dressed in matching ensembles constructed from fig leaves . . .

ADAM

So, what are we serving tonight, darling?

EVE

I thought I'd do that roast pig with the saffron and pistachio crust you like so much, with a light pomegranate sauce.

ADAM

Good choice. Which wines do you want with that?

EVE

Well, I went into Uruk this morning to the market and simply could not get my hands on that lovely red Mr. Hedjou makes. You know, the Egyptian who has a few vineyards in Libya? He told me that his vineyards are now almost all entirely covered in sand and he cannot even irrigate anymore. He's moved into a flat on the Nile and is going into date production.

ADAM

So he had no wines left?

EVE

No. And all my favorite pottery shops are closed. All I could get my hands on were these mass-produced copper bowls—hideous!

ADAM

My love, we are not in the protoliterate period for nothing! Ever since the Black Sea flooded again, and the Sahara is getting out of control, everyone is heading here, to our beautiful evergreen Garden of Eden. At least we're safe from the flood levels and the droughts.

EVE

I am so pleased that we landed up here and didn't follow your brothers east to settle on the Yellow River to grow that new thing and make that other new thing—what were they called again?

ADAM

Rice. Idiots—imagine using precious grains for something called "noodles" and not bread. What a waste. No future in noodles. No, nothing will get me to leave here.

EVE

Might I tempt you with a holiday to that new tropical resort on the Dead Sea that everyone is talking about?

ADAM

You mean Sodom? No thanks—I've heard that it's attracting the fast crowd. Better to enjoy the comforts of home, I say. So—I guess we'll have to make do with that Muscat from last year. Does it taste a bit flabby to you, by the way?

EVE

It will be lovely, don't worry. I can always add a few herbs and spices to jazz it up.

ADAM

What's for dessert, then?

EVE

Why, your favorite, of course. Apple pie!

Versions of this conversation may well have echoed throughout the eras. Tacitus and his peers surely bemoaned the loss of their favorite wines grown on the banks of Mount Vesuvius after its eruption in AD 79. Although not as much as they mourned the much greater loss of their drinking companion Pliny the Elder, the great Roman naturalist, when he left the sanctity of his villa on Capri to sail across the Bay of Naples and get a closer look. Wine regions have always been coming, going, and coming back again.

What we consider to be our invariable certainties today were once part of someone else's unimaginable scenario. While all the experts' attention has been consumed with the idea that humans are affecting the climate, we forget to consider how much the climate has always affected human society. Today, now, it is a mixture of both. An archaeologist attending the American Geophysical Union 2003 Annual General Meeting presented research suggesting that climate change was responsible for the collapse of society in the very cradle of civilization, ancient Mesopotamia (Megan Sever, "Mesopotamian Climate Change," *Geotimes*, 2004). There is evidence for a mass migration from the more temperate northern Mesopotamia to the arid southern region around 6400 BC. For the previous thousand years, people had been cultivating the arable land in northern Mesopotamia, using natural rainwater to supply their crops. Why did the people move from an area we knew had fertile farming to a more arid, difficult area? Ms. Sever quotes Harvey Weiss, an archaeologist at Yale University, explaining that one reason could be climate. There was an abrupt climate change event in 6400 BC, followed by a period of immense cooling and drought that persisted for the next two to three hundred years. Weiss believes that when the severe drought and cooling hit the region, there was no longer enough rainwater to sustain the agriculture in the north, and irrigation was not possible due to the topography, so these populations were left with two choices: to embrace a nomadic existence, or to leave, to migrate.

It was more arid there, but the society had learned to take advantage of the plain-level Tigris and Euphrates and create comprehensive irrigation systems. Irrigation farming is more labor-intensive than rain-fed irrigation, but this is better than no crop—remember our wine producers in present-day Languedoc-Roussillon. One added advantage of irrigation farming is higher yields (this is one of the reasons why it is a no-no in finewine production). But for food staples, it is a blessing. It meant people experienced food surpluses and thus were less preoccupied with the immediate need to nourish themselves. They could focus on "full-time crafts rather than relying exclusively on farming, thus giving rise to the first class-based society and the first cities," concludes Weiss. So we can say that adverse climatic conditions forced the advancement of our socialization skills.

Our ancestors have never stop moving—from the birthplace of humanity in the Great Rift Valley of East Africa at least sixty thousand years ago, to their crossing of the "gate of grief," the Bab el Mandeb strait that separates Africa from Arabia, to their explosion in "just 2,500 generations, a geological heartbeat, to the remotest habitable fringe of the globe . . . to our species' last new continental horizon, Tierra del Fuego in Southern Chile. We are living through the greatest mass migration our species has ever known" (Paul Salopek, "Out of Eden," *National Geographic*, December 2013). We now know that the Afar Desert, once fertile, experienced a drought that lasted (is lasting) thousands of years that may have trapped early humans in Africa, as travel would have been too risky. It would have been a climate shift of wetter periods that would have enabled the first periods of migration (Salopek).

So our Adam and Eve had to move to find a more hospitable environment. And their Garden of Eden would later be flooded by the waters of the gulf, when in 5000–4000 BC the Flandrian Transgression caused a sudden rise in sea level; it reached its present-day level about 4000 BC (Dora Jane Hamblin, "Has the Garden of Eden Been Located at Last?" *Smithsonian Magazine*, 1987). See what a bit of flooding can do? Responding to our immediate climatic environment is what shapes us, and forced adaptation can lead to stronger evolutionary traits and survival skills. Interestingly, I also read in chapter 2 of John Heise's *Akkadian Language* that "there was a new hot and dry period, starting around 500 BC, which hastened environmental changes (overgrazing and deforestation) [and] probably did contribute to weaken the Mesopotamian civilization and caused the 'center of civilization' to move to northern latitudes." So from northern Mesopotamia to southern and then back again. Here, clearly, we see that the populations were shifted back and forth over the millennia as the climate ebbed and flowed.

We know that drought ravaged the Mayan civilization, and archaeologists have also linked another, earlier climate event, about forty-two hundred years ago, to the collapse of the societies found between the Aegean and the Indus during the Early Bronze Age. But "it's perhaps too extreme to say that climate change caused all of the advanced society collapses, but it's also too extreme to say that climate change has had no effect. The challenge to us as paleoclimatologists is to develop much more detailed and well-dated records" (Peter deMenocal, Columbia University's Lamont-Doherty Earth Observatory, "Mesopotamian Climate Change").

This theory that climate change is what sets history on its course is called climate determinism, or environmental determinism. Hippocrates introduced the theory in his fifth-century AD *Airs, Waters and Places*. It contends, broadly, that climate influences the psychological disposition of different races, and critics have always suggested that determinism served to justify racism and imperialism. Again, a debate I do not wish to enter. But I would add that I never thought the EU was a good idea because of climate. Work ethics will not find a common denominator; they shouldn't attempt to. Why should they? We are all different. We are all products of our environment, and that should be allowed. Have you been to Sicily in August? Noto? Climbed those stairs of the Cattedrale? Can you blame them for taking a siesta from noon till sundown? No amount of icy limoncello can soothe that sort of hot. Northern European tribes had to work harder to survive. Every moment of their day was monopolized with the tasks of survival: finding food, building shelter, burning fuel. Meanwhile, in the south, fruit was falling from the trees and fish were jumping onto the shore. Simple.

I have strayed. What does all this have to do with wine? Well, mass migrations are not the sole preserve of the human species: Plants, too, like to travel. And the *Vitis vinifera* has always had a bit of the wanderlust.

THE *VITIS VINIFERA* . . .
BAGS PACKED AND READY TO GO!

When I first began writing this book, I was ready to espouse a long-held belief of mine: Because the *Vitis vinifera* is indigenous to Central Asia and the Mediterranean basin, any attempt to grow it outside of its home habitat would produce an inferior product. It is a cool climate plant and has no business in a valley floor in Napa or Maipo. I have always contended that while variations on a theme of Chardonnay (for example) should be accepted, for me all non-indigenous or non-Burgundian versions were not only "different" but "inferior." It's a rather conservative view that is not always appreciated. But would you rather eat an apple grown in Florida, or one grown in Washington State? That said, if I follow that thread to its logical conclusion, I would then have to pretty much discount all wine regions west of the Black Sea. Because, in a properly historical context, France is "New World." So I thought about it some more.

We are happy to say that Sauvignon Blanc, Sémillon, Cabernet Sauvignon, Petit Verdot, Merlot, and Cabernet Franc are "indigenous" to Bordeaux. But they are not. They were brought there. I once overheard a very heated discussion between two colleagues at a wine judging event in Bergamo, Italy. We were tasting only Merlots from the surrounding region. The debate was concerning whether or not Merlot was "indigenous" or "traditional" to the region, having been planted there over 150 years ago. The winner of the debate (if such a debate can be won) argued that 150 years constitutes "traditional," and anything much older as "indigenous." But as climates changed, and as humans traveled toward what would become Europe, they took the wine grape with them and cultivated it in their new homes . . . in Greece, in Italy, in France. The wine grapes acclimated and adapted. Which means that *Vitis vinifera* should do the same in the New World. So I thought about it some more.

This is what I came up with. If *Vitis vinifera* adapted so well in Europe, it was because the climate was so similar to that of its original home. Also, it was given a very long time to settle in. Which means that my stance still holds water. Sending the wine grape off to considerably warmer and drier climates than to which it was accustomed, without so much as a bottle of sunscreen, and telling it to move in, unpack, and immediately prepare a fresh, elegant, and sophisticated five-star meal in a new kitchen is ludicrous. And if that kitchen has no running water, has a cupboard full of tin cans, and is equipped with only a microwave, then it is nigh impossible. That's the analogy. I like to think of it another way, too. A traditional species imposes itself upon a new environment, while an indigenous one flourishes in the natural conditions of its home environment. A being should not have to "force" itself to survive in a climate by manipulating it. If you have to live in a place where water is fed via canals, residing in an air-conditioned house, driving an air-conditioned car to an air-conditioned mall to buy imported, vacuum-packed vegetables or vegetables grown under poly-tunnels or reliant upon irrigation, fertilization, and extreme mechanization, then isn't something wrong?

So I will go on record as saying that perhaps we should have foreseen that some of the struggling New World wine regions would have their *Vitis vinifera* visit end in tears, as welcoming as they were as hosts. But now even my "perfect model" of Chardonnay, back in its "home" in Chablis, is lacking its luster. Home isn't as hospitable as it once was. Its climate, too, is changing. The Burgundian model ever-increasingly resembles the warmer climate models. And wine producers all over Italy and France are harking back to other grape varieties that used to be grown in their region, varieties even more indigenous than the internationally known varieties of today. Bordeaux used to (legally) grow Carménère, and many others. Is it time to change . . . or to move?

WHY CLIMATOLOGISTS LOVE
THE WINE INDUSTRY

Climatologists love the wine industry. They love the record keeping and the harvest reports. They love the plethora of references found in our classic literature. They love the tomes of wine tasting notes. More particularly, they love grapes. Grapes are considered an "indicator crop." It is a research-er's dream: constant, reliable, and subject to meticulously detailed record keeping. Even my humble personal research brought me to wine producers who have compiled harvest records that date back 150 years and more, in many cases. All this provides a wealth of information. "Because wines are constantly being tasted and rated for quality, wine grapes are a particularly good indicator of changes that are probably affecting other crops in the same areas" (Gregory Jones, Southern Oregon University). Those wine grapes grown in their European indigenous climates are the "control group" against which all others, from other climates, are measured. They are the yardstick of quality. Don't forget: Wine is grapes and grapes are fruit and fruit is farming. Wine production is glorified gardening—the same golden principles apply. The greatest winemaker I ever met (in my view) once told me once that he considers himself a "farmer," nothing more.

In his article "Climate Change and Wine," Gregory Jones is another who refers to the grape as the agricultural "canary in the coal mine" in refer-ence to the impact that climate change will have on wine production. He writes: "Climate is a pervasive factor in the success of all agricultural systems, influencing whether a crop is suitable to a given region, largely controlling crop production and quality, and ultimately driving economic sustainability. Climate's influence on agribusiness is never more evident than with viticul-ture and wine production, where climate is arguably the most critical as-pect in ripening fruit to optimum characteristics to produce a given wine style" (Noah S. Diffenbaugh et al., *Climate Adaptation Wedges: A Case Study of Premium Wine in the Western United States*, February 2011).

When we say that the grape acts as the canary in the coal mine, we mean that it is the perfect indicator fruit, our climatic crystal ball. Grapes, and more specifically the wine grape, are the most sensitive of fruit to climatic changes and influences, where temperature is considered the most important of all variables; the equation also includes *terroir*, soil, winemaking practices, and more. It has the narrowest band of climatic suitability, and any shifts within this band are reflected in the grape's ripening ability and the resulting wine's quality. The narrower this band becomes, the higher the quality of the wine; the looser this band—that is, the farther the grape finds itself from its center, whether toward the northern, cooler parameters of the band or toward the southern, warmer parameters—the more we see quality lessen. So sensitive is the grape to extremes of either heat or cold that it tells what impacts we can expect on other fruits and crops. This is a source of invaluable information and one that climatologists relish.

A Brief Outline of the Earth's Climate History

Hadean Era (4.5–3.8 billion years ago)

The Big Bang. Earth a molten mass. Cooled down enough for rain, which created oceans.

Archean Era (3.8–2.5 billion years ago)

The first landmasses appeared, and life developed. The atmosphere was still toxic, mainly consisting of ammonia and methane. The sun was only about 75 percent as bright as today, but the earth was not completely covered by ice. The planet was even warmer than today. Scientists assume that a greenhouse effect caused by high methane levels in the atmosphere prevented freezing.

Proterozoic Era (2.5 billion–500 million years ago)

Atmospheric oxygen levels slowly started to increase, fueled by photosynthetic algae. This probably wiped out a huge portion of the earth's anaerobic inhabitants, thus the era's label as the "oxygen catastrophe." Evidence points to several periods of glaciations, but little is known about their causes. Scientists assume ice covered almost the entire planet.

Paleozoic Era (500–250 million years ago)

The planet's core finally cooled down to a level comparable to today. Volcanic eruptions became rarer. Earth generally saw a lot of glacial activity because the two primary supercontinents drifted across the North and South Poles, receiving little solar energy. The rapid development of photosynthetic organisms made the air more breathable. During the so-called Cambrian Explosion, life in the oceans evolved rapidly from simple to complex forms.

Mesozoic Era (250–65 million years ago)

Earth's climate was mostly dry and highly seasonal, with large temperature differences. When another one of the supercontinents broke up into smaller units, more land came in contact with the oceans. Humidity therefore increased, and the climate became warmer and wetter. Temperatures were about 18°F (10°C) higher than today and about the same everywhere, from the poles to the equator.

Cenozoic Era (65–2.7 million years ago)

Around sixty-five million years ago, an asteroid as big as the Isle of Wight smashed into what is today Mexico. The impact left a huge crater, and a layer of dust blanketed the earth. Dinosaurs became extinct; mammals took over. About fifty-five million years ago, the planet experienced sudden warming. Methane bubbles, previously hidden beneath ice sheets, burst and further pushed the warming trend—theorists point to a massive volcanic explosion. In just twenty thousand years, global temperatures shot up 9 to 14°F (5–8°C). The earth was ice-free. Some thirty-three million years ago, the first glaciers started to form in Antarctica, finally heralding an era of cooling. Ice sheets grew; sea levels fell. Instead of tropical forests, grasslands started to develop. This cooling trend culminated in the ice ages of the Pleistocene epoch.

Pleistocene Epoch (2.7 million–12,000 years ago)

The earth's orbit changed and less solar energy hit the surface, which triggered the ice ages. The planet had been almost ice-free for billions of years. Now the advancing ice covered large parts of North America, Europe, and Asia. Cold glacial periods took turns with short warming spells, when glaciers retreated. It is estimated that there were between twenty-five and thirty ice ages and interglacials. During the coldest periods, average temperatures were about 7 to 9°F (4–5°C) lower than today. In between ice ages, some lesser peaks of temperature occurred a number of times, especially around 125,000 years ago. Temperatures then may have been about 2 to 4°F (1°–2°C) warmer than today, and sea levels were sixteen to twenty-six feet higher than today. This is a rise sufficient to inundate most of the world's coastal cities and was caused by orbital cycles.

Holocene Epoch (12,000 years ago–present)

Around twelve thousand years ago, the last ice age ended and earth experienced a warming phase—the Climatic Optimum. But while other epochs were defined solely by natural and geological phenomena, the Holocene has become the period of human influence on biosphere and climate. Temperatures in the Northern Hemisphere were warmer than average during the summers, and the tropic and areas of the Southern Hemisphere were colder than average. Between about 3000 and 2200 BC the once-lush reaches of North Africa and

Arabia turned to desert. Drought destroyed the Indus Valley people. In 2000 BC, wetter, stormier times came. This happened again in 1000 BC, and then it became warm and dry again: Greece and Rome flourished between 500 BC and AD 400. Droughts returned, and crops in North Africa failed, while forest and grass vanished from Lebanon and Galilee. Then came the cold and wet Dark Ages. The climate re-warmed a bit between AD 800 and 1000, when the earth experienced the Medieval Warm Period; the Vikings exploded out of Scandinavia and settled in a greener Greenland, Iceland, and even Vinland. Wine grapes were cultivated in England, farther north than even now. By 1300, temperatures started to drop again. The direct sea route from Iceland to Greenland became impassable. Greenlanders died to the last person by the late 1400s. The Baltic Sea froze over in 1422–23. From the sixteenth to the nineteenth centuries, we saw the "Little Ice Age" when glaciers all over the world expanded. The climate was too cold for satisfying crop yields, and famines, plague, and revolts struck Europe's peasant populations. In the American Revolution, guns were dragged across the river to Staten Island. The year 1816 was "the year without summer." Snow fell and freezes ruined crops in July and August. This was due to the 1815 eruption of the vol-cano Tambora in the East Indies. Things changed abruptly in the twentieth century. The earth's surface temperature rose on average by 1.3°F (0.7°C). The increase occurred in two periods—from ap-proximately 1910 to 1944 and from 1978 to 1998. And this is the part that incites the debates: This increase is attributed to the ef-fects of the Industrial Revolution, or increased human population and activities (Bettina Fachinger, "The Temperature of Europe During the Holocene Reconstructed from Pollen Data," *Quaternary Science Review*, 2003; "What's Happening to Our Climate?," *National Geographic*, November 1976; the 2013 IPCC Report).

WHY GRAPES GROW WHERE THEY DO

As stated above, the grapevine tolerates a narrow climatic range—and that range is shifting, which is what is confusing us so. With the continued rise of global temperatures, the *Vitis vinifera* will have to travel as much as a thou-sand kilometers in latitude outside of what was once considered its indig-enous habitat and its previously traditional climatic range. It has to do so to survive. It requires a growing season long enough to complete the fruiting cycle and a period of winter cold to force the vines into dormancy. It needs certain amounts of light, warmth, and water. All vines belong to the family of climbing forest plants adapted to seek daylight through trees. It needs the light to promote photosynthesis, although cloudy light will do. And it only grows within a narrow temperature band. Vines start to grow at 50°F (10°C), do best between 59 and 77°F (15–25°C), but slow down above 77°F. At around 82°F (28°C), moisture will evaporate from the plant's leaves faster than it can be drawn from the ground. Growth and photosynthesis will cease.

Wine grapes can grow and produce wines outside these norms, but they will have either insufficient or overabundant yields. Climates at either end of the spectrum do not attain the typicity or quality of a fine wine. If you have a region with average temperatures below 55°F (13°C), you are most likely limited to hybrids or very early-ripening cultivars that do not necessarily have large-scale commercial appeal. This is the case with German and Austrian wines, as well as English wines. It is only recently that *Vitis vinifera* varieties have been able to reach maturity in the UK. If you have a region with an average growing season temperature above 70°F (21°C), it would be mostly limited to fortified wines, table grapes, and raisins.

The grapevine thrives best in wet, humid conditions. But this does not suit producers, because this is conducive to large yields and fruit that easily rots or splits. Vines will seek moisture by sending down roots as deep as possible. Deep roots ensure against drought and also reach the mineral sources. And deep roots can be encouraged by planting the plants closely together (planting density) to thereby encourage competition among them and force them to delve deeper. When planting density is low, this means that the plants can spread out and their roots can take up as much space as they like on the ground surface. When they are irrigated and heavily fertilized, this is further impetus to become lazy and avoid having to dig for nutrients. And just for measure: 8,000 to 10,000 plants per hectare (4050 per acre) is average for France (allowing for the width of a horse), while 454 vines per acre is the traditional average for California (allowing for the width of a tractor). But this is a vast topic. More about its relevance to climate change later.

The world's best vineyard sites are located on hills. The Romans, and even the Greeks and Etruscans before them, had figured that out. Slopes allow good drainage, avoiding "wet feet"; they shelter the vines from wind, and can maximize the vines' exposure to the sun. In the Northern Hemisphere, classic wine regions were rarely sited at high altitudes because this meant the effect of latitude was modified and the areas would have been too cold or windswept. However, this is changing in Italy and Spain, where the latitudes provide hot summers that are now getting hotter, and so everyone is heading to the higher altitudes. In the Southern Hemisphere, we see the same thing: New

Zealand's Central Otago, for example is their southernmost highest-altitude region. Valleys and flatlands are not conducive to quality grape growing: They make it too easy for irrigation and machine-harvesting, and generally are composed of more fertile soils, which drives yields up, and high yields do not equate to fine wine. As with children, if you spoil the vine, you spoil the wine.

THE EARTH'S CLIMATE CATEGORIES

ARISTOTLE

All attempts to classify the earth's climate zones started upon Aristotle's rather limited view of the world. His understandably oversimplistic view divided the world into three parts. The Torrid Zone runs from the Tropic of Cancer (23.5°) in the north, through the equator (0°), to the Tropic of Capricorn (23.5°) in the south; the Frigid Zone comprises the area north of the Arctic Circle (66.5° North) and south of the Antarctic Circle (66.5° South); and the two Temperate Zones, found between the Tropics and the Arctic and Antarctic Circles). Aristotle thought that both the Torrid and the Frigid Zones were uninhabitable, and that, of course, the zone in which he lived was the only one that would allow the advancement of our species.

Aristotle's Model:

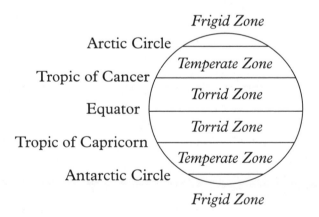

WLADIMIR KÖPPEN

In 1884, the Russian German climatologist Wladimir Köppen (1846–1940) published his climate classification system, based loosely on Aristotle's but introducing a far more sophisticated system of climate region types and sub-types. He later updated it in 1918 and in 1936, and eventually collaborated with Rudolf Geiger. He thought that native vegetation is the best expression of climate and so used combined average annual and monthly temperatures and precipitation and the seasonality of precipitation. There has been much criticism of Köppen's system, but it is still the one used. What is interesting is that all this data is now shifting. If you visit the Köppen research site (http://koeppen-geiger.vu-wien.ac.at.), you can see a digitalized updated version of the map (and others) based on data from the Climatic Research Unit (CRU) of the University of East Anglia and the Global Precipitation Climatology Centre (GPCC) at the German Weather Service.

KEY: Precipitation

W = desert
S = steppe
f = fully humid
s = summer dry
w = winter dry
m = monsoonal

KEY: Temperature

h = hot arid
k = cold arid
a = hot summer
b = warm summer
c = cool summer
d = extremely continental
F = polar frost
T = polar tundra

So:
Af = Group A (tropical), subgroup f (fully humid)
Csb = Group C (Mediterranean, warm temperate), summer dry and warm summer

GROUP A Characterized by constant high temperature. All 12 months have average temperatures of 64°F (18°C) or higher.	Tropical humid (equatorial)	Af	Tropical wet	No dry season	NW Pacific coast of South and Central Asia; Indonesia, Malaysia
		Am	Tropical monsoonal	Short dry season; heavy monsoonal rains in other months	Cairns, Queensland, Australia; Miami, Florida
		Aw	Tropical savanna	Winter dry season	Darwin, Caracas, Mumbai, Bangkok

(continued)

GROUP B Characterized by annual precipitation less than a threshold value set equal to the potential evapotranspiration.	Dry/arid	BWh	Subtropical desert	Low-latitude desert	Khartoum, Sudan (BWhw), Alexandria, Egypt
		BSh	Subtropical steppe	Low-latitude dry	Niamey, Niger; Libya; Priorat, Spain; Mendoza, Argentina
		BWk	Mid-latitude desert	Mid-latitude desert	Omnogovi Province, Mongolia
		BSk	Mid-latitude steppe	Mid-latitude dry	Xining Qinghai, China; Montana, Wyoming, southern Idaho, parts of Colorado, Utah, Nebraska, parts of North and South Dakota
GROUP C Climates with an average temperature above 50°F (10°C) in the warmest months and a coldest-month average between 27 and 64°F (-3 and 18° C).	Mild mid-latitude (warm temperate)	Csa	Mediterranean	Mild with dry, hot summer	Tel Aviv, Israel; Seville, Spain; Perth, Western Australia; southern France and parts of Italy; Coonawarra, Australia
		Csb	Mediterranean	Mild with dry, warm summer	Porto, Portugal; San Francisco, USA; Napa Valley, California; Pacific NW; southern Chile; west-central Argentina
		Cfa	Humid subtropical	Mild with no dry season, hot summer	São Paulo, Brazil; Barolo, Italy
		Cwa	Humid subtropical	Mild with dry winter, hot summer	Bengbu, Anhui, China

(continued)

		Cfb	Marine west coast and "highlands"	Mild with no dry season, warm summer	Burgundy, Bordeaux, Piedmont, Austria; Auckland, New Zealand; London; Melbourne, Victoria; Vancouver
		Cfc	Marine west coast	Mild with no dry season, cool summer	North Vancouver, Canada; Reykjavik, Iceland
		Csc	Subalpine	Dry-summer maritime	Balmaceda, Chile; Bohemia Mountain, Oregon
GROUP D Climates with an average above 50° F (10°C) in the warmest months and a coldest-months average below 27°F (-3°C).	Severe mid-latitude	Dfa	Humid continental	Humid with severe winter, no dry season, hot summer. Tend to be found above 40° N latitude.	Chicago, Illinois; Romania, Ukraine, Moldova, northern Japan; southern England; Middle Atlantic states; Midwestern states; southern Ontario, Canada
		Dfb	Humid continental	Humid with severe winter, no dry season, warm summer	Eastern Europe/ Black Sea basin; Helsinki, Finland; Kiev, Ukraine; Fargo, North Dakota; Yellowstone Park; Buffalo, Finger Lakes, New York; Montreal, Quebec, Niagara, Canada; Germany; eastern Central Europe; southern Alps of New Zealand; Snowy Mountains of South New Wales; southern Andes of Chile and Argentina

(continued)

		Dwa	Humid continental	Humid with severe, dry winter, hot summer	Pyongyang, North Korea; northeastern China near Yellow Sea; Korean Peninsula
		Dwb	Humid continental	Humid with severe, dry winter, warm summer	Vladivostok, Russia
		Dfc	Subarctic	Severe winter, no dry season, cool summer	Murmansk, Russia; Yellowknife, Northwest Territories; Crater Lake, Oregon
		Dfd	Subarctic	Severe, very cold winter, no dry season, cool summer	Eastern Siberia
		Dwc	Subarctic	Severe, dry winter, cool summer	Mohe County, China
		Dwd	Subarctic	Severe, very cold and dry winter, cool summer	
		Dsb	Warm summer continental	Higher altitudes and latitudes in North America	Flagstaff, Arizona; South Lake Tahoe, California
GROUP E Climates with average temperatures below 50°F (10°C) in all months of the year.	Polar	ET	Tundra	Polar tundra, no true summer	Mount Fuji, Japan; Mount Washington, New Hampshire; Jotunheimen, Norway (and occasionally Pic du Midi de Bigorre in French Pyrénées)
		EF	Ice cap	Perennial ice	Antarctica; inner Greenland

A. J. WINKLER

Then along came A. J. Winkler and Maynard Amerine at UC Davis and their heat summation method. Knowing that the grapevine does not grow below 50°F (10°C) (from Köppen's Group C), Winkler measured the days in the growing region (April 1 to October 31 in the Northern Hemisphere; October 1 to April 30 in the Southern Hemisphere) and assigned degree days according to the amount that the day's average temperature exceeds this temperature: one degree day per degree Fahrenheit over 50°F. It is a measurement of warmth, of sunshine: the cooler regions (Region I) have the lowest number of degree days. He allowed cloudy-day sunshine to count, which growers criticized—it changes the results, and growers only really want direct-sunlight days. He then matched the grape varieties to their best climate regions. He devised the system for California and the rest of the United States and superimposed the rest of the world's wine regions onto his measurements.

Region	Degree Days	MJT °C	Grape Varieties	Wine Region Examples
I	1,111– 1,390	<19.8	Early-ripening varieties achieve high quality: Pinot Noir, Riesling, Chardonnay, Gewürztraminer, Pinot Grigio, Sauvignon Blanc	Chablis (all of the Côte d'Or), Friuli, Rhine, Willamette Valley, Tasmania, Champagne, Marlborough, Finger Lakes
II	1,391– 1,670	19.9– 21.3	Most early and midseason table wine varieties will produce good-quality wines with light to medium body and good balance: Cabernet Sauvignon, Chardonnay, Merlot, Semillon, Syrah	Bordeaux, Alsace, Yarra Valley, Frankland River
III	1,671– 1,940	21.4– 22.8	A favorable climate for high production of standard to good-quality full-bodied dry to sweet table wines: Grenache, Barbera, Tempranillo, Syrah	Rhône, Clare Valley, Lower Hunter, Rioja, Piedmont
IV	1,941– 2,220	22.9– 24.3	Favorable for high production, but table wine quality will be acceptable at best: Carignan, Cinsault, Mourvèdre, Tempranillo	McLaren Vale, Upper Hunter, Langhorne Creek, Montpelier, Languedoc-Roussillon, Spain
V	>2,220	>24.3	Typically makes low-quality bulk table wines or fortified wines: Primitivo, Nero d'Avola, Palomino, Fiano	Greek Islands, Jerez, Sicily, Sardinia, North Africa

Dr. Andrew Pirie, "Defining Cool Climate," Stratford's Brave New World seminar, London, September 2007.

When these measurements were taken (in the 1970s), Champagne, Chablis, and the Loire, all with degree days between 1,710 and 1,980, were cooler than California's coolest region, the Monterey Peninsula, at 2,160 to 2,340. The "cool" areas of Monterey along with the Santa Clara and Livermore Valleys are all appreciably warmer than the Côte d'Or and Alsace. The average of the degree-day range for California's Napa Valley was the same (2,475) as the Médoc.

I viewed this as a sort of attempt to legitimize California's wine regions as a member of the fine wine club by virtue of climatic association. Like, if Napa, the "new girl," hangs out with Bordeaux, the "cool girl," long enough, people will think they are close friends because they are so alike. Other people have criticized the system by saying that it does not take into account all the microclimates in California. This is true; still, no existing climatic system takes into full account the microclimates anywhere. The plethora of Burgundy's microclimates is the perfect case in point. One point on which most agree is that it is temperature that is the predominant factor in a grapevine's development.

This is a good place to suggest that it will be the microclimates that will provide our future surprises and our exceptions to the rules. When any system measures only temperature or precipitation, it will be missing out on the other vital components that combine to create a true climate. This is what we call *terroir*.

TERROIR: THE MAGIC TRILOGY OF SOIL, CLIMATE, AND GRAPE VARIETY

Terroir is one of the most evocative and much-debated terms in the wine lexicon. What is it? It is not just "dirt," as one flippant journalist from Sydney once dared to quip out loud while sitting in a tasting room in Meursault. Let's begin with the word *soil:* its richness in fertilizing elements (this affects a plant's vigor); its structure (compact, rocky, or muddy); its mineral composition (granite, chalk, or limestone); its color (red soils warm faster in spring); and its topographic situation (on a hill, in a valley, on a plain?). Grapevines need soil that will store moisture, drain excess water, and force the vines to grow deep roots to search for nourishment, thereby developing character and strength. The best soils are not too fertile—rich soil produces very ordinary wines.

But there is more to *terroir* than soil. *Terroir* is not a place, but a happening—a combination of circumstances. Either it happens or it doesn't. It cannot be manufactured, contrived, or forced. *Terroir* encompasses a site's topography, hydrology, geology, sunlight, and climate. These are a lot of variables to get right. And all of this has to be in place before humans even get involved.

Is *terroir* terribly important? Well, there is good wine, and there is great wine. Good wine can be grown and made almost everywhere—which is why we will not have to fear a global wine shortage. Even as we are adapting to climate change and our regions and their output are shifting, there will always be enough good wine to go around. Great wines, however, are more particular, and it is these wines that will be the hardest hit. Great wines need *terroir*—that magic trilogy of soil, climate, and grape variety. This must remain the incontestable and objective indication of a wine's quality.

Traditionally, *terroir* was only ever associated with the Old World European wine regions. This is a very divisive debate: the variability of the Old World versus the climatic reliability of the New World (which equaled "boring and easy"). Let's shift the focus of this debate. Indeed, climate changes are forcing us to do so. Let's not pitch Old World traditions against New World technicians. The two worlds are quickly and confusingly converging—climate-wise, not *terroir*-wise, sorry. I have never tasted a fine wine from a New World country that rivaled one of its counterparts in Europe. I am sorry if that makes me unpopular, but that is what I have gleaned from my vast experiences, and I am not alone in thinking it, although I may be alone in writing it.

Do you remember the producer in Emilia-Romagna during the blazing harvest having to face irrigation, machine-harvesting, and other new practices? These were things that New World winemakers had to embrace from their inception, not practices that were slowly imposed upon them by their climate changing. The New World is working toward fine wine within the context of climatic limitations. But now it feels as though the climate is moving faster than producers are and pulling the rug out from under their feet, undoing all their hard work and erasing their previous potential. I wish it were not so. I must add here that when I speak of fine wines, I, personally, am not referring only to those famous Grands Crus with the big price tags. For me, the definition of *fine wine* has to be enlarged to include those wines that are traditional and *terroir*-driven, the wines that speak to me from whence they came. I meet these sorts of wine often on my travels: they are distinctive, personable, and seductive.

What we can say is that both worlds are capable of producing international, homogeneous, correctly-made wines to please the mass market. France and Italy are able to produce the most divine nectar on earth, but we all know that they have made their mark on the other end of the spectrum as well, pouring forth unspeakably banal wines—again, usually from their least distinctive, warmest *terroir*s.

We know that *terroir* affects the taste of wine. Just do a taste test of Chardonnay from Auxerre to Saint Véran. All white wines in Burgundy are issued from Chardonnay (and a trace of Aligoté), and all reds, from Pinot Noir.

And as there are not a dozen different ways to elaborate a wine, especially if you studied at the Michel Rolland School of Wine, then the only variable in this equation is *terroir*. Or take the microclimate of Les Bessards on the Rhône's Hermitage Hill. This is sandy gravel on granite soil showing a wine with a spicy leathery nose and a tannic palate that will age well, as opposed to Le Méal, which has stony soils on hills and terraces, and which will show wines that are very fruity, delicate, and elegant, with a more rounded tannin structure and thus a shorter guard.

We also know which grape varieties do better in which soils. Gamay, for example: In the granite soils of Beaujolais it produces a fine, agreeable wine. But when planted a few miles north, in the chalky soils beloved by Pinot Noir and Chardonnay, the Gamay becomes heavy and loses its perfumes. The Old World has had plenty of time to experiment with these matings: Chardonnay to Chablis, Tempranillo to Rioja, Sangiovese to Chianti, Nebbiolo to Piedmont, Riesling to Mosel, Tourigo Naçional to Duoro, Sauvignon Blanc to Sancerre, and so on. So much so that these matings are set in stone, somewhere in an office in Brussels. They were created with good intentions, but they are going to have to change so that the Old World winemakers are free to replant and experiment in the same way as do their New World cousins: They will have to start the process all over again.

The crux of the *terroir* debate is that those who do own vineyards in the established, traditional appellation are able to sell their wines at higher prices, and those who do not, cannot. The New World has been beavering away at creating its own perfect matches; some pretty violent disputes have erupted in California as to where to draw these magic boundaries. Tim Mondavi once said to me that Californian appellation demarcations are openly acknowledged to be more about marketing than about *terroir*.

But I do believe that it is an inherent compulsive need of a winemaker to want to get the most from the land, to make the best wine possible. Perhaps this is a naive belief in some cases.

On my first trip to South Africa, I was shocked to visit wineries that were growing up to ten different grape varieties: parcels of Chardonnay next to Viognier and Sauvignon Blanc; Muscat next to Grenache; Grenache next to Pinot Noir. I called them the "fruit salad" wineries. When I asked about why they grew so many different varieties from such different climates and soils, the answer was always: "Well, you see, my land is so topographically diverse. I have so many various soil types and microclimates that I can accommodate them all." Come on. Finally, one brave man told me what I had guessed: that wineries were hedging their bets as to market trends and wanted to be ready and poised to respond with a Pinot Grigio one year or a Merlot the next. Others, he said, were genuinely just trying to get to know their land and needed to find what did best where through a process of elimination, and that although they were starting with a huge range, their goal was to hone it down to one or two grapes—those that best match their site and soil. We forget that this is why it is called the New World. They are still experimenting. We should allow for the eventuality of future great fine wines being born of these soils . . . if only climate change were not changing their variables so quickly.

Ten *Terroirs*

1. Meursault, Côte de Beaune, Burgundy: For its buttery Chardonnay—petit déjeuner in a glass

2. Chablis (unoaked), Burgundy: For its crisp, mineral Chardonnay

3. Richebourg, Côte d'Or, Burgundy: For its voluptuous, opulently animal expression of Pinot Noir

4. Margaux, Bordeaux: For its sophisticated and perfumed Cabernet Sauvignon–dominated claret

5. Guebwiller, Kitterle, Alsace: For its Pinot Gris SGN that flirts between steely Riesling and unctuous Gewürztraminer

6. Fiddletown, Sierra Foothills, California: For its oldest plantings of hearty, spicy Zinfandel

7. Maipo Valley, Chile: For its deep-colored, full-bodied Carmenère

8. Barolo, Piedmont, Italy: For its powerfully dangerous and brooding Nebbiolo

9. Tokay-Hegyalja, Hungary: For its dizzyingly majestic Tokaji Aszú Essencia (mostly Furmint & Hárslevelü)

10. Falerno del Massico, Campania, Italy: For its revival of the ancient Aglianico and Piedrosso

IS CONTINENTAL THE "NEW" MEDITERRANEAN? WINE'S TRADITIONAL CLIMATE CATEGORIES

In Europe, the wine climate model, as was Winkler's, is loosely based on Köppen's Group C: a sort of subset classification. They are: Mediterranean, maritime, and continental. Again, the *Vitis vinifera* does its best work when grown between the thirtieth and fiftieth parallels in both hemispheres . . . for a little while longer, that is. This is what is shifting. Köppen's Group C includes climates with an average temperature above 50°F (10°C) in the warmest months and an average of between27 and 64°F (-3 and 18°C) in the coolest months. Already, the regions he included in his categories outside his Group C are taking steps into viticulture, and some of the regions in his Group C will eventually no longer be able to sustain viticulture.

Mediterranean	Has two subsets: Hot Summer (Csa) and Warm Summer (Csb). See Köppen chart. Hot Summer is what we consider to be "typical": average monthly temperatures higher than 71.6°F (22°C) during the warmest months and an average in the coldest months of 64 to 27°F (18 to -3°C) or between 64 and 32°F (18 and 0°C). Hot, dry summers and mild, wet winters. High summer temperatures can be cooled by nearby large bodies of water. Growing seasons are long and of moderate to warm temperatures. Little seasonal change with temperatures in the winter generally warmer than those of maritime and continental climates. During the grapevine-growing season, there is very little rainfall (with most precipitation occurring in the winter months), which increases the risk of drought. Climate change is seeing more and more Csb sites move into Csa classification.	Tuscany, rest of Italy, Greece, Spain, Israel, Lebanon, Southern Rhône, Languedoc-Roussillon, Provence, Portugal (ex. Douro), Slovenia, Croatia, California, Western Australia
Maritime	Köppen's Cfb, broadly. Regions that are close to large bodies of water (oceans, estuaries, inland seas). The "middle" range between Mediterranean and Continental. Also has a long growing season but often suffers from excessive rain and humidity, which bring disease. Clear seasonal changes, but not as erratic. Warm summers, rather than hot, and cool rather than cold winters. (not anymore!)	Bordeaux, Muscadet, Willamette Valley, Long Island, most of New Zealand, Southern Chile
Continental	Köppen's Cfb, Dfb. Hot summers and moderately cold winters. Acute seasonal changes throughout the growing season. Hot temperatures during the summer and periodic ice and snow in winter. Usually inland. Big dips in temperature between day and night. Winter and spring have risks of hail.	Columbia Valley, Burgundy, Rioja, Piedmont, Northern Rhône, Douro Valley, Loire Valley, Austria, Hungary, Romania (the entire Black Sea basin), Russia, Turkey, Columbia Valley, Canada, Mendoza

Pinot Noir: Finding a New Attitude in New Latitudes

With the boiling vintages in Europe, such as in 2003 (the one that prompted Sarkozy to allow winemakers in Languedoc to irrigate), Burgundy's Pinot Noir risks getting flabbier and more alcoholic. The Burgundy identity is fast being erased and starting to resemble the warmer New World models. This will work for a while—it's an easy wine to sell. But for me, jammy, flabby caricatures are boring. So where to look now?

The Pinot Noir is mistakenly considered an easily adaptable grape. It is true that its clones morph with very little prompting, and it seems happy enough with enough sun to ripen it, but it is, in fact, a very hard grape to get right. There is a difference between just ripening a grape and capturing its varietal essence. The balance is delicate, and climate and soil are critical.

> An example of cool-climate suitability is found with Pinot Noir . . . which is typically grown in regions that span from cool to lower intermediate climates with growing seasons that range from roughly 57.0–61.5°F. The coolest of these is the Tamar Valley of Tasmania, whereas the warmest is the Russian River Valley of California. Across this 4.5°F climate niche, Pinot Noir produces the broad style for which is it known, with the cooler zones producing lighter, elegant wines and the warmer zones producing more full-bodied, fruit-driven wines. Although Pinot Noir can be grown outside the 57.0–61.5°F growing season average temperature bounds, it typically is unripe or overripe and readily loses its typicity (Dr. Gregory Jones).

Closer to home, I have been following the Pinot Nero in the Alto Adige region, nestled in the Italian/Austrian Alps; Alsatian Pinot Noir, previously metallic and light, seems to be gaining in body; and Austria's Blauburgunder is delicious, as is Germany's. Forget Riesling; Germany is really undergoing a red wine revolution.

> Germany has experienced an average of 1°C warming during the last century, but this is not distributed evenly throughout the year. For example, the average temperature for August is up 2°C on a century ago, and this is the most vital period

for tannin formation in pinot noir, a process accelerated by higher temperatures. This is an example of how a little bit of warming can change the entire ball game for certain grape varieties. Now, pinot noir will regularly ripen in vineyard sites with southerly exposure as far north as the Mosel and Ahr (at a latitude of 50° 30' North). In warmer years winemakers have to carefully watch picking dates to avoid excessive alcohol levels (Stuart Pigott, "German Pinot Noir: Insipid? No, Inspiring," 2013).

More recently, I tasted a series from New Zealand's Central Otago, the southernmost region. The region is so new that most of the plantings are only ten years old. Yet there were a few there that were more typically young Burgundian than Burgundies: which undid every single one of the prejudices to which I have been clinging for most of my career. So, there is much to look forward to.

WHEN IT ALL GOES RIGHT

The weather during a growing season will determine how many grapes are harvested and in what condition they will be once harvested. Sun, snow, rain, frost, heat . . . they are all needed, but at the right amounts and at the right times. Here is an ideal harvest profile.

WINTER

After harvest, which in the Northern Hemisphere is usually during September/October, the vine closes down for the winter. A bit of frost and snow is beneficial now, as it kills any remaining pests and diseases. Heavy rain is also welcome at this point as it replenishes the soil and fills its water reserves. This is also when the major pruning is done. The Guyot system, for example, involves pruning back the vine to just one or two canes and then training these along wires. This will keep the vine from overproducing. Best if done by hand.

SPRING

The vine comes back to life with the first rays of sunshine. The recently pruned canes glisten with syrupy ears of sap. Then, near February, little furry buds develop on the knob of each cane. At the first signs of spring, as the temperature warms, these buds begin to open: budburst (*débourrement*). Tender little cabbage-like heads of green leaves appear, in the middle of which are nestled minuscule bunches of grapes. This stage is slow moving, but speeds up as soon as the air reaches a constant warm temperature. This is the most delicate period of the vine's life—the time when it is most vulnerable to sudden spring frosts (usually in May), pests, and diseases. Warmth and some light rain are needed now to ensure continued growth. April is also usually the period during which to plant new vines, if necessary.

SUMMER

In June, when the air temperature reaches about 70°F (20°C), the buds will flower: *floraison*. Early flowering is a sign of a good vintage. But if the weather is too cold during flowering, the grapes may not develop properly. During this period, the vine needs long days of sunshine while it pollinates. The berry starts to take shape, called setting. If all goes well, it quickly grows larger and heavier. Once it reaches the height of its development, it changes from its green color to either a transparent yellow (if it's a white grape) or a deep violet (if a red grape). Called *véraison*, this change usually happens in the summer. The grapes fill up with sugar and water, becoming fully mature, and hang around, awaiting harvest. In July, heavy "green pruning" should be practiced—thinning out the green grapes to reduce the crop size.

AUTUMN

This period can make or break a vintage. One wishes to harvest under warm, clear skies.

SUMMARY: THE PERFECT SEASON

- A winter sufficiently humid to reconstitute the plant's reserves yet cool enough to avoid an early budburst.
- A spring that is warm and dry, with no rain, to allow a rapid budburst, which allows a rapid vegetative maturation, which then protects plants from parasites and unwanted frost.
- A summer that is even warmer and drier, but not too hot, to allow a good *véraison* and a good concentration of sugars, both of which help the plant to continue fighting off parasites. An excessively hot and short summer will create false grape maturation.

WHEN IT ALL GOES WRONG

The weather conditions that pose the greatest threat to a maturing vine are the winter and spring frosts, along with any hail, rain, extreme cold, or extreme heat or drought. A winter freeze is especially dangerous if the temperature drops below 5°F (-15°C). This can cut a harvest in half and leave the remaining crop damaged. It can even kill the vine. This used to be an unusual event—no longer! Note that a light layer of snow can actually help to insulate the vines from stagnant freezing air. This is why winemakers often spray their vines with water, hoping that it will freeze into a sort of protective capsule.

A spring frost is even more dangerous because the plant has just burst open and is very vulnerable. Any damage at this point means fewer flowers . . . and fewer grapes. Such a frost often sneaks in during clear, calm nights, when temperatures may drop to 30°F (-1°C). If it happens on an evening when there are clouds in the sky and a little wind, the danger is minimized; the wind creates a bit of turbulence, drawing heat from the ground, and then the clouds retain this heat. Winemakers spend May on nightly "frost watch." And many a night, as a guest at a château, have I been woken by the warning bell at 2 AM and run outside in my nightdress to help light the *bougies* in an attempt to warm up the vines and ward off the impending frost. It works!

Hail is most dangerous after *floraison*, when the newly formed grapes can be shattered and split. This is literally known as shatter.

Rain during *floraison* can cause *millerandage*, which is a partial form of shatter and prohibits the plant from blooming. The flowers are pollinated but not yet impregnated, which means that they won't acquire pups, but will remain small and green. Uneven development can occur within the same bunch. Too much rain also means too much humidity and a greater chance of rot and mildew. Rain during harvest, once the grapes are mature, means swollen watery grapes, and diluted wines. This is why knowing when to harvest is an art form. A perfect season can be ruined by an unexpected last-minute September rain, which will ruin the health and concentration of the grapes, and therefore its quality and aging potential.

The same is true of dryness and heat. A winter that is too warm will allow a precocious budburst, which will leave the plant too vulnerable to spring frosts. A sudden heat wave in August or early September can block the maturation process and create overly tannic, alcoholic wines. This is what is happening more often. The heat raises the sugar levels, but arrests the maturation process—so we get green tannins and unripe but alcoholic grapes.

THE GENIE IN THE BOTTLE

It is so that we can determine a wine's future that we examine a wine's growing season—its childhood, its breeding. Some might say that wine professionals are frustrated futurists by trade. For those who have been trained in the dark art of divination, it is a skill that is becoming mired in conflicting codes and clues. The harvest report is the code; the first taste of the nascent wine, the first clue.

It is the appellation, or more specifically the *terroir*, of a wine, in conjunction with the climatic events of its harvest year, its vintage, that ultimately determines its extraction and density, and therefore its aging potential. There are no shortcuts. It all starts at the dawn of the New Year, with the first ray of sunshine. Again, every drop of rain or dose of sun can alter the route from bud to bottle. Warm climates shorten the growing season, which ripens the fruit more quickly. This shorter "hang time" can mean that the fruit may ripen, but not the tannins, and we get green tannins. The longer and cooler the hang time, the better. Slow-ripening fruit becomes more complex and fully mature. A shorter growing season translates into wines that have a very upfront *première bouche*, or first attack with good primary extractions, but can be lacking in acidity to hold the structure together.

WHY, WHAT, AND WHEN IS MATURATION?

To make a wine that is so complex and full of matter that it needs a long time before it reveals its true beauty and balance—this is the entire point of winemaking. Only certain grapes and certain soils possess the ability to produce a wine that will age and improve. The story of a wine only really begins when it is placed in that bottle. This is the magic. To sit and wait and see what happens . . .

The maturation of a wine is in function of its composition, its origin (*terroir*), and its vintage . . . what happened to it in the vineyard. No two bottles from the same Bordeaux château, of two different years, will develop and mature in the same number of years. This is why the more we know of what goes into a wine and how it is made, the better we understand how and why it ages.

Almost 75 percent of wines produced today are meant to be consumed young. There is a fallacy that all French wines, and even all Bordeaux, merit long cellaring. The truth is that the majority of French wine styles are meant to be drunk two to three years after bottling. Even the top growth producers are making fine wines that drink better when young in order to compete with the New World models and to fit into the lifestyle of the wine consumers and their trends. Who cellars wine anymore?

No one knows for certain what happens during the aging process. We know that there is an olfactive evolution. The nose a wine has when it is young is called the aromas; once mature and oak-aged, these same odors become more complex, turning into the wine's bouquet. During bottle aging, red wines depose little plaques and grains of coloring agents and other molecules, which bond and fall to the bottom of the bottle. The heavier clusters settle faster and sooner than the smaller ones, which need years to settle. As these coloring agents settle in the bottles, the intensity of the wine's color diminishes, becoming more and more reddish brick, and finally yellowish, as the anthocyanins, or coloring agents, in the tannin soften and diminish while the tannins do.

When we say that the tannins soften, we mean that the wine has reached a point of balance in its development curve. The harsh edge is taken off the wine, and the wine's texture is smoother and less drying.

In general, the maturing of a wine is divided into four categories of phenomena: chemical, physical, biological, and physical-chemical. Chemical maturation includes the oxidation of the polyphenols, the alcohol, the sugars, and the organic acids. There is also the conversion of alcohol into formaldehyde, acetate, and ester. And finally, the hydrolysis of polysaccharides and glucose. Physically, there is the insolubilization of salts, the release of gases, the evaporation of volatile substances, and the dissolution of tannins. Biologically we observe the malolactic fermentation, which softens the malo acids, and a sort of fungus-like enzymatic change for those wines that were oak-aged. And finally (as if all this didn't already make your mouth water), there are the physical-chemical changes such as oxy-reduction, polymerization (the uniting of single molecules of the same substance to produce larger molecules, or "mers"), and the formation and flaking of colloids (substances that cannot crystallize and become solid, such as amidon or gelatin).

Polymerization progresses continually as the wine ages so that tannic wines for long aging become gradually harder and more tannic before reaching a peak where they are more tannic then when they were in barrel. Then the slope starts a gradual decline. The extra-large molecules lose their ability to combine with other proteins, and their astringency diminishes. At the same time they are combining with other components in the wine, becoming insoluble and precipitating to form the characteristic deposit. At this point the wine is in its mature, mellow phase and is softer, richer, and rounder: This is maturity. If kept too long, the increasingly large polymers gather strength once more, a sort of final wind, and become dry and astringent again. This is compounded by the fact that the wine is also losing its fruit and gaining volatile acidity. It is drying out, or dying.

But for a wine to mature in this way, it has to have all the components in place before it starts. We are waiting for all its components to fall into balance with one another. A good wine is one that is balanced and harmonious. The acids, alcohols, tannins, and fruits have all blended into one, creating a personality, a character. A less good wine will never fall into place, because the fruit will die, or the tannins will fade away into nothing instead of softening and holding the wine together, or the acids, the backbone and life of the wine, may become flabby or disappear. The greatest threat to balance now is the heat driving up the alcohol levels, which then dominates the wine and drives down the acidity, which makes a wine taste sweet and flabby. When tasting a young wine, we look for all the elements and try to determine how well the wine is made and how it will mature. This is becoming an increasingly more difficult exercise, as the alcohol gives the young wine a sweet palate and masks the other attributes. You used to taste a young wine and get a mouth full of tannins, but you could detect the good acidity and extracts, and knew that the tannins would balance out. Now, with the alcohol dominating everything and erasing the varietal character, it is hard to imagine that it won't dominate the rest of the wine and leave it forever unbalanced.

Certain wines will reach maturity sooner than others, and not every wine will have the same length of maturity. In general, the duration of a wine's ideal maturation period is the length it needed to reach it. If a 1982 Latour or 1982 Gruaud-Larose will reach its apogee in 1997, then we can hope to be able to enjoy our bottles for fifteen years. But then we can expect to continue drinking it for years more.

The great "vins de guard" have been known to age to fifty, sixty years and more. If you look at the excerpts from Michael Broadbent's *Great Vintage Book* in the next chapter, you will see that he has spent his career tasting Bordeaux that were decades old. The vintages since the mid-1980s may not do as well, and certainly those from the 2000 decade will not enjoy such a long life. Is this due to the consumer's preference for younger, fruitier wines? To the technological changes viticulturists have been making in the winery? To climate change ramping up the alcohol and unbalancing the wines? And does it matter? Do we have to drink old wines? Is the enjoyment of an aged Richebourg any less than that of a fresh, vibrant Alto Adige Pinot Nero? Is it time for us to redefine the requirements of great wine? Everything we have ever known is now being challenged and changed.

A Brief Climate Change Q and A

I know that I said that I did not want or need to go into the mechanics of climate change. But I did speak to a UK aficionado specializing in eco-architecture, Michael Campbell Bell, ask some questions, and request that the responses be easy for me to understand. Here are the responses. Do with them as you will.

What is the difference between climate change and global warming?

Global warming is just one of the effects of climate change—a subset. *Climate change* is the better term as it includes all extremes of all weather events and the rising sea levels. The earth is warming faster than it has in the past thousand years, but in fact, some areas may cool.

What is the jet stream?

There is more than one jet stream. But the jet stream we talk about is high in the atmosphere and it moves from east to west or vice versa. However, we are most affected as it drifts between north to south like a waving band according to the temperature on either side of this band—pushing and pulling it about. Much of its recent activity is probably due to the high temperature in the Arctic.

What is the Gulf Stream and what is it doing?

The Gulf Stream is a current of water that rises up from the Caribbean, all the way up the East Coast of the United States, and then fans out across the Atlantic. When it reaches the Campbell ice spit—which is a large chunk of ice that projects out from Greenland—masses of cold fresh water come pouring into the North Atlantic. It drives the current down to a lower depth—which then returns at a low depth back to the Caribbean. It then recirculates—warming and cooling all the time. As Greenland's ice melts, the effect of this is drastically diminished and so the cold water is only dropping down to a much shallower depth as it heads back to the Caribbean. So you have cooler water cooling the upper warm water, or cooling the Gulf Stream. This slows the Gulf Stream down, and it could eventually stop. This won't be the first time.

What happens if it stops?

The UK would adopt our "correct" latitude climate. Fifty degrees North without the Gulf Stream is like Siberia and Canada. So we would get cold winters and hot summers.

What about CO_2?

As we pump CO_2 into the atmosphere, the air temperature rises. There is a big debate about whether temperate proceeds or follows CO_2. As CO_2 rises, ten to twenty years behind that, temperature rises. So the massive rise today will cause higher temperatures in the future—but as to when, that is a bit of a mystery. The CO_2 is forming a blanket around the earth and creating a greenhouse effect. Worse than this, as the Arctic ice melts, methane is released from the permafrost, causing homes built on this ground to tilt at alarming angles. The world trade in ivory is now focused on woolly mammoths. Their corpses are coming to the surface—defrosting from their icy graves.

Reports have been that the Arctic has gone up 7°F (4°C) and is melting. But then I read that this year it did not shrink as much as last year. Is it melting, or not?

This year it did not shrink as much as last year. The year 2012 was the worst so far, and 2013 was a bit better. These fluctuations will start and stop, but will continue upward—climate change is not linear.

What is the relevance of the ocean water heating?

The effect of the ocean heating up is driving marine life farther north, to cooler water. In the Southern Hemisphere, Antarctica is actually getting colder. The current "pause" in global warming is attributed to massive amounts of heat being absorbed into the Pacific Ocean.

If everything is so "stop-and-start," how do we track these trends?

This is why we have the greatest computer in the world trying to figure this out. There are other layers to the situation, other factors. For example, solar activity is decreasing. There are fewer sunspots. At the current rate there will be no sunspots by 2026, and this could signify the beginning of a mini ice age *within* the warming: the trend within the trend within the trend. We've got the world heating up, the Atlantic cooling down while also warming, with sunspots decreasing. Every factor, every climatic variable, has its own cycles and trends and they are all playing off one another and triggering one another.

So what does this mean for Europe?

It should continue warming, but no one is sure what this ice age will do: Will it be centered only in Europe? Or elsewhere? Anywhere affected by the Gulf Stream is cooling, while Europe is also warming. We don't know which trend will dominate.

Why are English wine harvests warming, but getting wetter, while during the same harvest in southern Italy the grapes die on the vines from drought?

Because the UK is in the Gulf Stream—Europe less so. They are more affected by the position of the jet stream. The jet stream is normally north of the UK during the summer, but recently it has been dropping down to central France (hence the hail and erratic harvests). When positioned north of the jet stream we tend to get bad weather, and when south we tend to get good weather. So we get rain and Italy gets drought: the two extreme interpretations of "bad" and "good." These cycles are getting more and more erratic and disparate. The movement of the jet stream is key here, and we understand it the least of all, apparently. We forget that New York City is the same latitude as St. Tropez! So while New York is feeling the effects of global warming, we are feeling the effects of global warming and the Gulf Stream slowing at the same time, so our rise in temperatures will be slower than in New York, or the United States. The Hadley Centre computer shows the UK as being the least affected by global warming (but not necessarily by climate change) by the year 2100. And overlaid over all this is the lack of sunspots.

So still, why all the bad weather if the earth is warming?

This has to do with the difference between "weather" and "climate." Even in a warming climate, we will get individual weather systems that will bring bad weather. Remember: cycles within cycles and trends within trends. Over the past one hundred years, warming has meant fewer frosts and more heat waves in many parts of the world. The amount of rainfall is getting heavier in some countries in terms of volume per downpour.

What do sunspots do?

They are coronal mass ejections—electromagnetic pulses. Good or bad? A lack of them means that the sun is going through a cooling period. But these eruptions start from the core of the sun, I think, and it takes thousands of years to come to the middle of the sun and then to the surface of the sun, to cool.

**I read an article from a 1970s volume of *National Geographic*
that said that then, the world was *cooling*, not warming.
So have we heated up since then?**

Because the effect of CO_2 has massively increased since the 1970s—at
the time then, global warming had not had a lot of effect. In the '50s,
'60s and '70s, we didn't really notice it. But all the while, we did have
the eleven-year cycle of the sunspots. We had a low level back then and
it was a cooling period. Again, a cycle within a larger cycle. Layers.

I am still confused. Are we cooling or heating?

Both. The world will get a bit cooler. It is heating up with CO_2, so
there will be a lull in the atmospheric heating as the oceans absorb
the heat while the solar activity calms down. As we insulate ourselves
with more CO_2, we increase global warming. And as the CO_2 and
methane continue to be produced, we continue to insulate ourselves.
The fire is stoked for the night.

Will the UK get warmer and wetter before cooling again?

As it gets warmer the Gulf Stream slows down because the ice cap
is melting. When the Gulf Stream slows, the UK assumes its more
logical latitude characteristics. This means colder winters and hot
summers, but all the while, "warmer" colder winters and "cooler"
hotter summers . . . and lots of rain. Get it?

**What should European wine producers do with all this hail
and rain?**

Start covering them with nets, I say! All crops are suffering. Here in
the UK, oak and beech trees no longer thrive. We are being advised
to plant ash, hornbeam, and whitebeam. Lovelock said that if the
CO_2 keeps rising at the same rate it is now, the only time CO_2 has
reached the level we have today was due to intense volcanic activity.
This is the only time it has gotten to that level without such volcanic activity. And if it carries on, we will go back to alligators in the
Arctic—but when? We'll be in biospheres, living artificially. But we
will have plenty of solar power for electricity. The way people live
now in places like Phoenix, Arizona, will be how you live everywhere.
Hot, sunny countries should only be using solar electricity. Their
using air-conditioning is doing us all in. It is a self-perpetuating cycle.
Use the sun! To build an uninsulated, non-solar-powered building in
a hot country should be a criminal offense. We have to force building
authorities to work responsibly. We all do.

2

WHAT IS HAPPENING
IN THE VINEYARDS

Harvest Report: 121 BC

Among the remaining wines no kind was particularly famous, but the year of the consulship of Lucius Opimius, when the tribune Gaius Gracchus was assassinated for stirring up the common people with seditions, was renowned for the excellence of its vintages of all kinds—the weather was so fine and bright (they call it the "boiling" of the grape) thanks to the power of the sun, in the 633rd year 121 BC from the birth of the city; and wines of that year still survive, having kept for nearly 200 years, though they have now been reduced to the consistency of honey with a rough flavour, for such in fact is the nature of wines in their old age; and it would not be possible to drink them neat or to counteract them with water, as their over-ripeness predominates even to the point of bitterness, but with a very small admixture they serve as a seasoning for improving all other wines.

—Pliny the Elder, *The Natural History*, Book 14, s. 55

A t this writing, the last of the 2013 harvest reports are still coming in from a few countries. Of course, the Southern Hemisphere is done and bottled by now. The Northern Hemisphere's sweet wines will still be on the vines, and the ice wines won't be harvested until January. That is, if there is any in 2013. The warmer temperatures of last year meant ice wine production in North America and in Europe was dramatically reduced, and some producers were not able to make any at all. Temperatures have to drop to a minimum of 19°F (-7°C), and in places like Germany's Mosel, they didn't drop below 46°F (8°C).

Harvest reports are the code we use to determine a wine's quality, both actual and future, assuming it is a wine meant to be capable of aging (not all are). Of course, the harvest report does not take into account the entire equation. A lot of things affect a wine's taste. It is a long journey from bud to bottle, and even when Mother Nature does her part, humans certainly can make a mess of things.

The decisions taken by the winemaker as to where to plant, what to plant, how to treat the vines, and how to harvest the wines, if taken incorrectly, can undo a good growing season. Wherever and whenever possible, you want hills, not valleys; soil that is healthy but not too fertile; high planting density; low intervention and not total manipulation; ruthless pruning and not vine "pissing"; hand-harvesting not machine-harvesting; rainfall and not too much irrigation; indigenous yeasts and not packs of powdered, selected yeasts; indigenous grapes not international grapes; balance and not over-extraction; low yields not high yields . . .

Good wines come from stressed vines. Great wines, from great struggle. Vines that are given all that they want and need, in excess, become spoiled and produce wines without character or personality. This equation is rooted in factual trial and error. If the qualitative measures above are not taken by a winemaker, it is usually for one or both of two reasons: his or her commercial agenda, and his or her climate.

A GOOD UPBRINGING ENSURES A GOOD TASTE

Wine is simply the result of the partial or entire alcoholic fermentation of fresh grapes or the juice of fresh grapes. It tastes like the things it contains, therefore the predominant flavors of wine come from the skin, or just beneath the skin, of the grapes used to make it. Grapes are composed of: water (between 75 and 90 percent), alcohols, acids, polyphenols (or coloring agents), sugars (fructose and glucose), carbon dioxide, and aromatic components (chemical). In addition to these, there are all the components of wine that we cannot see, smell, or taste, such as vitamins, proteins, amino acids, and so on. Each of these plays an important role in the taste of a wine. I have said this before, but like human beings, a wine's taste is going to depend a great deal on both its origins and its upbringing. In fact, the French use this very word *upbringing*, or *élevage*, when describing a wine's early life.

The vinification method, how the wine was treated before being bottled, and how long it has been left to mature in the bottle, all add to its unique flavor combination. Will it be a flash-in-the-pan sort of wine or will it mature nicely and develop even more character as it grows old? All in all, has it been disciplined or spoiled rotten? A spoiled wine, like a spoiled child, will be lazy, brash, and superficial. Again, a wine with character will behave more subtly and will reveal its strength and personality as you get to know it.

WHAT DOES WINE TASTE LIKE?

The taste elements of both wine and food are the same: sweetness, acidity, bitterness, and astringency. The sweetness in the taste of wine comes from the fructose and glucose—types of sugar—in the grape and from the alcohol produced during the fermentation. The alcohol is not sweet in itself; it just underlines those components that are, and helps to counteract the acidity and tannins, making the wine appear sweeter. The acidity comes from the tartaric and malic acids in the grapes. Malic acid is green-tasting and can make wine very bitter, which is why the fermentation processes chosen by the winemaker are so important in achieving a balance—they will affect the malic acid levels in the wine. Bitterness and astringency come from the tannins in the grapes. There are very few tannins in white wines, so most discussion of tannins is reserved for the reds.

THE TASTE OF ACIDITY

The sour-tasting substances in wine are the acids: tartaric, malic, and citric are in the grape; succinic, lactic, and acetic result from fermentation. The total acidity of a wine depends upon whether the growing season was too cold—in which case the grapes are too acidic and underripe—or too hot—in which case the grapes become overripe and lack acids. White wines generally have more acidity than reds. It is the amount of acid that is important: too little and the wine is bland and flabby, too much and it is vinegary. The right amount of acid, in balance with the wine's other components, makes the wine look and taste crisp, clean, and lively, as well as ensuring longevity.

Tartaric acid is unique to grapes and to wine and represents one-third to one-quarter of the total acid composition of wine. It is the strongest acid and it strongly influences the pH of a wine. The pH measures the concentration of hydrogen ions, which for wine means its dryness. The lower the pH, the safer the wine is from diseases and from oxidation, and therefore the greater its aging potential. The tartaric acid content decreases as the grape ripens, then varies depending upon the harvest weather conditions.

Malic acid is found in every part of the grapevine. It is the most fragile of the acids, which allows for its easy transformation into lactic acids (called malolactic fermentation), which diminishes considerably the overall acidity of a wine. The hotter the year's weather, the faster the acid decreases during the ripening process, which is why there is more of it when the weather has been cooler. All red wines are allowed to go through a complete malolactic fermentation. White wines can go without, go partially through, or go entirely through this second fermentation. It depends upon the juice's initial acid and sugar levels and the style of wine desired. If a winemaker wants a crisp and green white wine, then the malolactic fermentation is halted or not even allowed to begin. For a buttery, smooth white, the malolactic fermentation is permitted for longer or until its completion.

Acids give a wine its shine or brilliance, especially the tartaric acids, which renew the wine's color. It is the presence of malic acid that often gives a wine an apple smell, and in the mouth we can sense the amount of acids by the irritation of our gums and the inside of the mouth.

THE TASTE OF TANNINS

Tannins are the most important of the three types of polyphenols, or coloring agents, which give a wine texture rather than taste. Tannins that are condensed are present in the grape, and those that are exogene are procured from the wood during barrel aging. In the stalks, skins, and pips there are tannins that are released during fermentation and pressing, giving the wine its specific character and contributing to its aging capacity. Storing, or aging, the wine in new oak introduces additional tannins, which are transferred from the wood's fibers. These are more common in red wine than in white. Tannins obtained from the oak barrels can improve a wine's aging potential, complement its texture, and fill it out, but only if the wine itself has a solid backbone of acids, fruit extracts, and condensed tannins. Oak aging cannot replace raw materials that are lacking in the first place.

The wine's red color fades as the anthocyanins, another tannin, diminish with age. The more mature a wine, the more yellow or brown the *disc*, or surface, becomes. The combination of these coloring agents can give the wine a sour taste and a drying sensation in the mouth. This is called astringency. Different tannins from different-aged wines and wines from various regions will have distinct sorts of astringency. For example, young Bordeaux will have tough, astringent tannins, while old Bordeaux will have velvet-soft tannins. The more tannin is present in a wine's youth, the more it makes the sides of your mouth pucker and the longer it will take to mature. Don't buy a recent-vintage Bordeaux Premier Cru and expect it to go down like velvet; it is meant to age and mature in the bottle. This is why bottle-aging is so important, as is discussed later.

THE TASTE OF SWEETNESS

There are three major sweet-tasting substances in wine: the sugars and poly-alcohols, both originating in the grape, and alcohols from fermentation. Each style of wine has a different level of sugar depending upon the grape's maturation when harvested. Sweet wines contain several dozen grams of sugar per liter, while a dry white wine normally contains less than two grams per liter. The sugars, along with the alcohols, give the wine body and are visible because of the legs formed on the side of the glass. The sugars don't really have any odor but could eventually contribute to the overall expression of the wine. And it is the sugar that gives the wine its sweet taste, its fatness, and its unctuousness.

Alcohol is an important element in wine—it gives it great-looking legs. It is produced during fermentation when enzymes created by the yeasts change the sugar of the grape juice into alcohol, carbon dioxide, and heat. It is the proportion of alcohol to glycerine that determines the limpidity, or the body, of a wine, which we observe as legs or tears. More alcohol and the wine is thinner, thus the legs run down the side of the glass more quickly. The more glycerine present, the thicker the wine will be and thus the more slowly the legs will drip down the side of the glass. It is also primarily the amount of ethyl alcohol that will determine the sweetness of the wine.

THE TASTE OF BUBBLES

Of course, bubbles don't have a taste, per se, but they do release the wine's perfumes and give it a lovely texture. Carbon dioxide is the principal product, along with ethyl alcohol, of the alcoholic fermentation. It is present in both still wines and effervescent wines. If the wine is effervescent, carbon dioxide manifests itself as bubbles. If anything, you can detect an acidic taste edge while it pricks and tickles the tongue.

THE WEATHER AFFECTS THE TASTE OF THE WINE

The weather during any given growing season will change the taste of the re-sulting wine. The more heat during the growing season, the more sugars, and thus the more alcohols and the more sweetness. The more rainfall at harvest, the more dilution. Any extreme of weather leads to "no taste," whether it is diluted or so alcoholic that all nuances are negated. This is true of any crop, and particularly of fruit. Farmers all over the world have these same decisions to make and are having this same conversation. But it is "temperature that is key to all aspects of viticulture," assures John Gladstone (*Wine Terroir and Climate Change*). "The evidence is now clear that, with only minor other influences, it alone controls vine phenology, i.e. the vine's rate of physiological development through bud break to flowering, setting, *véraison*, and finally fruit ripeness."

In the UK, a country where high temperatures and sunshine are not an issue, the increasingly wet weather has another effect. One of Jeremy Vine's London radio programs in January 2013 had for its daily topic: "Have you noticed that your fruit and veg don't taste as good this year?" He had consumers and farmers from all over the UK calling to share their stories of their bad harvests. Rain. Rain. Rain. The potatoes weren't fluffy, the tomatoes weren't sweet and red, the fruit had no taste or crunch, the green beans rotted. The question came up as to whether farmers should give up on waiting for a decent harvest and start growing under poly-tunnels, to which then another stream of farmers called in to say that they were already doing that. So the discussion then moved to the comparative taste of those fruits and vegetables that didn't taste like themselves because of poor weather conditions, and the taste of those that were grown in a tunnel. The debate became heated.

The farmers against the poly-tunnels argued that the weather was only partly to blame. They said that the crops that did the best were those that are indigenous to the UK, like blackberries. They had the DNA strong enough and best-adapted to suit the varying weather. They added that the strains and varieties being planted by farmers—chosen because they are either fast-maturing (so they can get in more than one crop in a year) and/or high-yield (so they can get as much crop as possible)—are the problem. The decision to take shortcuts meant a crop might not survive the weather. The farmers who defended the poly-tunnels suggested that it was better to eat a ripe fruit, albeit a little bland and tasteless, than not any fruit at all. They said that market forces had made these decisions for them.

TASTING THE HEAT—WHERE'S THE BITE?

Important in the understanding of the warmer climate changing the taste of wine is the analogy of how the warmer climate changes the taste of other fruit: for wine is issued from grapes, which are fruit. And we want our fruit to be fresh, lively, and refreshing. Let's go back to Eden and take apples as an example. We want those to be crunchy and firm. A forty-year study of Japanese apple orchards has found that global warming is producing softer, sweeter, apples, writes Heidi Ledford ("Climate Change Threatens Crunchy, Tart Apples," August 2013). She quotes a study published in *Scientific Reports* on how changes in climate are affecting a huge variety of our staple foods, such as Fuji apples, sugar maple trees, and . . .wine grapes. "Climate changes are impacting the everyday lives of real people," says Christopher Field, an ecologist at the Carnegie Institution for Science in Stanford, California, who was not involved with the work. "It is not just an abstraction."

Fruit tree specialist Toshihiko Sugiura of the National Agriculture and Food Research Organization in Tsukuba, Japan, and his colleagues decided to study how warmer temperatures cause the apple trees to flower earlier and produce a riper, sweeter fruit. Ledford writes that they established that the "hardness and acidity of the apples had declined during that time, while their sweetness had increased." Mr. Sugiura says: "The changes may not be apparent to consumers because they took place so gradually. But if you could eat an average apple harvested 30 years before and an average apple harvested recently at the same time, you would really taste the difference."

Wine tasters have experienced exactly the same phenomena, and we have the tasting notes to prove it. Our favorite wines are tasting differently, and they are not as good. We miss that "bite." Jamie Goode of the wineanorak. com concurs:

> Relatively small changes in alcohol content can have quite a strong influence on how the other components of the wine are perceived. I find I don't really enjoy wines with higher alcohol as much because of the effects of the alcohol on the nose of the wine, and the bitter/sweet/salty character the alcohol lends to the palate . . . The other significant concern surrounds issues of "style" or "taste." Decisions about when to pick have quite an influence on how the wine will come out. In recent years there has been critical influence, largely from the USA, pushing red wines (in particular) towards a homogenized "international" style. I realize this statement could form the basis of a feature all on its own, but for now, I'm tempted just to say that red wines showing higher levels of ripe fruit, accompanied by softer tannins and plenty of new oak influence often get very high scores from the leading critics, whose ratings then influence sales, most notably in the USA where critical scores have a strong effect on sales. When grapes are picked late to achieve this style, and lots of new oak is employed in the élevage, the sense of place (or *terroir*) of a wine is often masked. Wines end up tasting similar no matter where they have come from.

He is absolutely correct and is not alone in his views: This is what is being said out in the field by both winemakers and wine tasters. Everywhere I travel, I am being hit with two stylistic camps of wine: the traditional and the international. Recent trips to Tuscany found us despairing over the popularity and prices of those alcoholic, in-your-face, boring Super-Tuscan monsters. And even among those produces who did stick to their guns and to their Sangiovese, there were inordinate numbers of sickly sweet, overboiled samples. If it's not Sangiovese, it's not Tuscan. And if it's over 14 percent alcohol, it's not wine, I say.

One afternoon during a tasting, held in a stunning thirteenth-century monastery, my colleagues and I were trying to figure out why this was so. We had compared notes and had all chosen the same estates as our favorites. We were a mixed group of nationalities: German, Italian, Russian, Spanish, Czech, and British. But we all shared the same concept of what a good wine was. Heartening. We had also all identified one estate as being particularly international and commercial and we decided to go back and give it another try, so scathing had been our notes. But wait . . . we could not get near the producer's table. There was a queue . . . of four Chinese journalists and three British supermarket buyers, all placing orders, filling in payment forms, all in a cloud of smiles and wildly happy international gesticulations. What does that tell us?

When I started out as a wine writer for *Vintage Magazine* in Paris, in the 1990s, the Bordeaux were at 12 and 12.5 percent alcohol. As the years have advanced, so have the alcohol contents. We are now drinking Bordeaux at 15 plus percent . Even the 1959 Bordeaux that was noted by Michael Broadbent as "the vintage of the century and one of the most massively constituted wines of the postwar era"—and which was a very hot, dry year—produced wines at decent alcohol levels. Château Latour was 11.6 percent (Andrew Jefford).

The hotter climate in Bordeaux is not entirely responsible for the increase. The higher alcohol levels everywhere are in fact due to, yes, rising temperatures, which means riper grapes with higher sugar levels, but also to improved viticulture (which means grapes are being picked in a riper state than they were before), and to stylistic changes, as winemakers have opted for later picking to produce that sweeter wine profile that marks the international style, as recounted above.

As my ex-editor at *Vintage*, Jacques Sallé, recently confirmed and explained to me, in Bordeaux it has also been most notably due to the viticultural practices adopted by *vignerons* since 1985. The increased heat helped the grapes reach a better maturity, from which then it is easier to get a better, more ripe extraction, which in turn produces wines that are more ample and generous. There are other factors that amplified this phenomenon: "Les scientifiques sont intervenus tout autant chez les pépiniéristes dans le but de produire un matériel végétal recherchant la meilleure synergie productivité/

qualité par sélection ou en faisant appel au génie génétique" (Jacques Sallé, 2013).

For example:

- Systematic anti-fungal treatments permit a better maturation of the grapes.
- Weed killers that destroy any and all other vegetation "competitors."
- The selection of rootstocks that are less productive than the SO4 (Sélection Oppenheim: *Vitis berlandieri* + *V. riparia*) means more concentrated grapes.
- Insecticides that destroy nearly all pests.
- The clonal selections.
- Excessive canopy management. Leaf removal allows for smaller, concentrated grapes.
- The "*tables de tri*": the hand selection of only the best and the ripest grapes.
- The control of fermentation temperatures.
- The selection of "killer" aromatic yeasts (OGN or selected).
- The concentration of the must by reverse osmosis.

All this, he continues, to create a style of wine that is more concentrated, overripe, and always with a touch of high alcohol to please US critics such as Robert Parker and thus obtain higher ratings and sell their wine at a higher price. The wines of the 2000 decade are heavy, inelegant, and boring. But— they are now great wines to drink when young, which is what they think the modern consumer wants. And as Jacques suggests, to be enjoyed with two ice cubes and a slice of orange. Absolutely.

He continues:

Au delà des ratings d'un Parker aujourd'hui retraité, au delà de tous les excès, il reste la terre, immuable, les vignes que l'homme a planté pour y produire le meilleur raisin dans le but d'élaborer le meilleur vin. Le vin n'est pas naturel, il est le produit de la volonté de l'homme pour son propre plaisir. Et si l'homme continue à brûler des matières fossiles, si le taux de CO_2 devient alarmant, s'il fait trop chaud, si les vignes donnent des raisins surmaturés et des vins trop concentrés, il faudra que l'homme change ses habitudes culinaires en adoptant une cuisine très épicée, et si le vin paraît encore trop dense, il suffira de lui ajouter deux cubes de glace et une tranche d'orange.

Translation

Beyond Parker's ratings, today, retired, and beyond all of the excesses, there remains the earth . . . immutable, the vines that man has planted so to produce the best grape in his desire to produce the best wine. Wine is not a natural product; it is the product of man's will, for his own pleasure. And if man continues to burn fossil fuels, if the CO_2 levels become increasingly alarming, if it gets hotter still, if the vines give us grapes that are over-mature and wines that are over-concentrated, it will be necessary for man to change his culinary habits, to adopt a cuisine that is much spicier. And if wine still seems too heavy and dense, then all there is to do is to add two ice cubes and a slice of orange.

So the answer to climate change is to start eating Mexican food!

The US wine critics' palate is very different from the European model. Prohibition is largely responsible for ruining, or at least retarding, the US wine heritage, as is the warmer climate of the US wine-growing regions. US and warm climate New World wines are more alcoholic, thus sweeter, so consumers prefer them and buy them. Understandably, the French got fed up with the mass consumer not understanding their wines, gave up, and copied the market leader. It was easy for Bordeaux to dumb down a bit with over-extraction and chaptalization. It would be impossible for California to catch up with Bordeaux. What is so ironic is that the French had to manipulate their wines to such a degree in order to achieve this international style—when little did they know Mother Nature was on the way to do it for them.

The Forming of the American Palate

How did this New World versus Old World palate divide come about? My theory is to blame it all on Prohibition. The American Prohibition is often romantically evoked by colorful images of speakeasies, bootleggers, and ax-wielding, saloon-smashing temperance ladies. But the reality was not so romantic. It was actually rather gruesome, really. And these social caricatures performed a great disservice to the country's wine industry. Prohibition (1919–33) shaped the way Americans now consume, appreciate, and produce wine. It was responsible for the "bulk wine" mentality of which the vestiges are still being fought by quality wine producers today. Prohibition was, in fact, a contradiction—there was nothing "dry" about it. Alcoholic consumption nearly doubled, both legal and illegal.

The early California wine industry, almost entirely consecrated to sacramental wines, took on another dimension with the advent of the California Gold Rush in 1849. Thousands of gold-hungry adventurers poured into the state, planning to strike it rich, and were forced to turn to relatively more mundane endeavors when their mines and dreams proved empty. Grape growing for the church was a popular backup plan. Then there were the Europeans who had already settled and were already established winemakers. The mix provided a strong foundation for a healthy wine industry.

National Prohibition was voted into law on January 16, 1919, just when the young California wine industry was at the peak of its glory. In 1920, more than seven hundred wineries were operating in California and wine acreage had reached three hundred thousand acres. California wines were being exported to England, Germany, and Canada, and had even won medals at international competitions.

Before Prohibition, Americans drank as the Europeans do, with food. This is logical, for the first Americans were largely European. The California wine industry was founded mostly by German, Swiss, and Italian immigrants, and wine, *good* wine, was a part of everyday life. So much so that when the Prohibitionists starting making noises as early as the mid-1800s, the wine trade did not take them seriously. Everyone was convinced that the Prohibitionists' target was hard liquor, and not wine. I think I even read somewhere that the original drafting of the legislation did not include wine in its definition of *alcoholic beverage;* it was later supporters who insisted that the definition be widened.

The biggest contradiction of Prohibition is that while over half of the existing wineries were closed, wine production and wine consumption increased. How did this happen? The Volstead Act declared the manufacture, sale, or transport of intoxicating liquors (wine, beer, and spirits) illegal. But there was an obscure provision in the act that permitted home winemakers to produce up to two hundred gallons annually of "non-intoxicating" cider and fruit juices for home use. And the sale of wine for sacramental purposes or religious rites was only authorized to a rabbi, minister, or priest. Now, you can just imagine the sudden religious verve that swept over the country! Everybody and his brother started a church or temple—or became a doctor. What a loophole. Remember the traveling "doctors" in their gypsy caravans and their "healing elixirs"?

The sad irony of Prohibition is that it did not destroy the California wine industry as it intended, but only retarded the advancement of the historic hardworking estates making the fine wine of the time. The wine producer and the market changed so drastically and quickly that there was no time, money, or demand for fine wine. Wine consumption increased as a result of the home winemaking boom. For those Americans who tasted their first wines during Prohibition, they would have been horrible, homemade, sweet, dark, boil-in-the-bathtub varieties. Prohibition changed the public's palate—its taste in wine. Whereas dry table wine had outsold sweet wine prior to Prohibition, sweet wine outsold dry by four to one following repeal. Americans had grown to prefer the high-sugar, high-alcohol wines of the era, and it was not until 1967 that this statistic was reversed. In fact, Prohibition reshaped (distorted) the American palate to such an extent that this explains our easy embrace of Coca-Cola and other sodas when introduced, and why Merlot (the sweetest of all the red grapes and used as the "sugar" in the Bordeaux blend) was, and still is, the most popular grape variety in the United States—the collective palate never recovered from this event.

Apart from the homemade concoctions, the rest of the wine being produced (legally) was equally bad. Why? Because the wineries still in existence survived Prohibition by supplying the numerous legitimate and illegitimate religious organizations, private households, and illegal speakeasies back east. This was such an enormous quantity demand that wineries were forced to tear out their quality grape varieties and replant with inferior varieties that produced higher yields—those with thicker skins that better endured the difficult transport conditions. Even worse, wine "bricks" became the norm. Four-by-eight-inch blocks of dried, compressed grapes were sold to the homemakers with the message: "Warning: Do not place this Wine Brick in a one-gallon crock, add sugar and water, cover, and let stand for seven days or else an illegal, alcoholic beverage will result!"

Companies sprang up selling these bricks. One company worthy of note was called Vine-Glo and sold "the pure juice of California wine-grapes, for home-use only." It was available in eight varieties: Port, Virginia Dare, Muscatel, Tokay, Sauternes, Riesling, Claret, and Burgundy. A customer would fill out and mail a coupon from a Vine-Glo advertisement, or place an order through a pharmacy. Then a five- or ten-gallon keg of juice would be delivered to the home and fermentation would be started by a company serviceman (Fruit Industries). In sixty days, the serviceman would return to bottle the wine, take the keg away, and start another batch if the client so wished. The five-gallon keg sold for $14.75, the ten-gallon for $24.40, and there was no charge for the bottling or the bottles.

Can you imagine a less auspicious start for a wine region than this? By the end of Prohibition, the industry was a mess. An entire generation of winemaking know-how had been wiped out, as well as the quality vineyards and their equipment. And because Prohibition was immediately followed by the Depression and the Second World War, it was very difficult for the industry to get back on its feet. It was not the time to reinvest, and so the bad wine kept pouring forth. Add to this the fact that with repeal, a new demand was made upon the wineries, which could not get the wines out fast enough and would often send wines that had been bottled before they had finished fermenting. Either the bottles turned to vinegar, or they exploded in the trucks and stores, thus quickly convincing consumers that what they had been concocting in their home kegs and bathtubs wasn't that bad after all.

There followed the era where enormous bulk quantities of generic jug wine hit the now legal consumer market. You know the stuff: Gallo "Burgundy" with not a drop of Pinot Noir in it—and the rest. Americans were now accustomed to bad wine: in jugs, in cartons, in screw-top jars. It didn't matter—we just wanted our white wine too cold and red wine too warm and sweet.

Today's heritage is a leftover of not only Prohibition, but also the later bulk wine days of the '60s and '70s and those of the tax haven and the "I am a dentist-with-a-chic-hobby" days of the '70s and '80s, when the California wine industry, as we know it, finally took off. California recovered from this false start. The current wine trade's biggest task had been retraining and reeducating the American palate.

Some of the wineries that survived Prohibition are still in operation today: Mirassou (1854), Sutter Home (1874), Beringer (1876), Simi (1876), Foppiano (1896), and Beaulieu (1900), to name a few. They have a rich history of sacrifice, struggle, perseverance, and innovation. It is important to note that most of the wineries either changed hands or knew some kind of interruption or upheaval during Prohibition. The fact that they have transformed themselves (once again) into international benchmarks of quality winemaking is nothing short of miraculous.

BORDEAUX BECOMES NAPA VALLEY

Bordeaux's top enologists have said that alcohol levels in Merlot have increased to such an extent that there is a very real danger the varietal character will be lost. (Panos Kakavialos, "Bordeaux 2010: Alcohol Is Threatening Bordeaux Style, Say Winemakers," 2010). Merlot is the sweetest of all the red Bordelaise varieties, meant to temper the tannins and comparative unripeness of the Cabernet Sauvignon, Cabernet Franc, Petit Verdot, and Malbec. It is the first variety to ripen, the first to be harvested. It is *already* the sugar in the Bordeaux recipe. It is now common to see Merlots exceed 15 percent alcohol. As the fifty or so leaders of the French wine and food industry wrote in their open letter to French president Sarkozy, "As flagships of our common cultural heritage, elegant and refined, French wines are today in danger. Marked by higher alcohol levels, over-sunned aromatic ranges and denser textures, our wines could lose their unique soul."

All varieties are now exhibiting higher alcohol levels, but this is more difficult to control with Merlot due to its nature. Kakavialos quotes Denis Dubourdieu, a top Bordeaux consultant winemaker, as saying, "A string of very good vintages since 2000 have meant there has been no need to chaptalise, so sugars are naturally there and that is a good thing." Well, no, this is a scary thing. It is legal in France to add sugar to encourage ripeness levels in cooler climates (chaptalization). Bordeaux always needed to do so. Then it was done for stylistic reasons, to get that New World upfront sweetness. So the fact that producers still create that style of wine *without* chaptalizing means that they have lost control of the harvests.

Dubourdieu adds that there is a tendency for properties on the Right Bank (where Merlot is the dominant variety, as opposed to the Cabernet Sauvignon–dominated Left Bank) to strive for greater concentration of sugars in the grapes by techniques such as leaf removal and late harvesting, sometimes even after the sweet wine harvests in October (Sauternes). "There is a race towards concentration, to please many critics," he says. "I have been a consultant for 30 years; I have spent the first 20 years telling people not to harvest too early; in the last 10 years I have told them not to harvest too late." Kakavialos goes on to quote Jean Claude Berrouet, formerly at Petrus and now winemaking director for various estates in Pomerol, as saying in his forty-year career, he has seen alcohol rise between 2 and 2.5 degrees.

Note: To harvest late is a good thing if the growing season is long and cool. This means that the grapes are reaching their maturity slowly and in a balanced manner. What Mr. Dubourdieu is saying here is that late harvesting when your grapes are already ripe and you are sitting on them wanting them to get overripe for "stylistic" reasons, or that late harvesting when the grape sugars are ripe but the rest of the fruit is not (waiting for phenolic ripeness), is not good.

Some like to put forth the argument that there have always been hot harvests, and that this has happened before. Well, that is not an argument. That is correct. But what these people are failing to realize is that the hot harvests are becoming hotter, and more frequent (and the wet harvests are getting wetter). Yes, there was '47, and '60. But considering the run of extremely hot vintages in the past decade (2000, 2003, 2005, 2009, 2010) . . . we cannot help but sit up and take notice. In the below excerpts from Michael Broadbent's *Great Vintage Book*, notice the numerous references to "scorching grapes," "drought," and "110°F" weather. But extreme heat was only ever representative of, say, one harvest in ten, or more. And because they were not using the same viticulture technique employed today to extract every drop of ripe fruit, the wines never reached those alcohol levels even with hot summers. This is no longer the case. NASA reports that 2012 was the 9th warmest in their analysis of global temperatures since 1880 and points out that while that seems unremarkable, what is important is the long-term trend, and the 10 hottest years in the 132-year record have all occurred since 1998; 9 of the 10 have occurred since 2002, in this decade. "What matters is, this decade is warmer than the last decade, and that decade was warmer than the decade before," says Gavin Schmidt, a climatologist at NASA's Goddard Institute for Space Studies. "The planet is warming."

After all is said and done, what we want from wine is for it to taste good. Forget investing, forget aging and cellaring, forget analyzing, dissecting, and classifying, we want enjoyment. We want wine to perform its purest function, to please and refresh us. Does bland, boring, sweet, and nondescript wine taste good? "Fine wine has detail, sense of place, nuance of variety and balance. Wines above 14% alcohol generally lack these pleasures. In the Yarra, and I suspect most regions, we lose a sense of detail and place with overripe characters," says Steve Webber of De Bortili, in Victoria's Yarra Valley.

As I recounted in the Introduction, so concerned about the run of hot vintages this decade, a group of French chefs, sommeliers, and châteaux issued a call to action, urging President Nicolas Sarkozy to put forth their case at the international climate conference in Copenhagen, or see generations of viticulture slowly die out as vineyards cross the Channel and head north. The signatories want the government to push for a global deal to cut industrialized countries' greenhouse gas emissions by 40 percent by 2020 and set up "solid aid mechanisms" for developing countries.

The signatories pointed to climate change with its heat waves, summer hailstorms, and new plant diseases as decimating the French vineyards. They said that if global temperatures rose by more than 2 percent before the end of the century, "our soil will not survive" and "wine will travel 1,000 kilometers beyond its traditional limits. We will have new wine-producing regions in zones where one doesn't normally cultivate vineyards like in Brittany and Normandy," says Jean-Pierre Chaban, a climatologist at France's National Institute for Scientific Research, in an accompanying online film. "It will spread to Great Britain. One can imagine vineyards in southern Sweden and Scotland."

RISING TEMPERATURES AND RIPENESS VERSUS MATURITY

We need to know the difference between ripeness and maturity in order to understand the damage that heat is inflicting on the growing cycle. The term *ripeness* usually refers to the grape while still on the vine, and *maturity* to the finished wine, a wine that is going to be cellared. That's easy enough. The hard part is defining, establishing, and achieving ripeness. Deciding when grapes are ripe is the most difficult and key decision a winemaker has to make. This is because once the grapes are picked, their quality is fixed in time, if you like, and their physical and chemical components are decided. Determining ripeness has become that much more difficult with the rise in temperatures. Overripe grapes have too much sugar, which creates high alcohol levels. High alcohol levels leave the wines unbalanced, and this compromises their journey to maturity.

The criterion for ripeness in grapes is the same as it is for all other fruit. The only difference is that ripeness in wine may vary according to the desired wine style: Sweet wine styles will require different degrees of ripeness measured in natural grape sugars at harvest. It is important to note that varying degrees of sugar content in these wines will not reflect in sweetness in the finished wine. Yes, riper grapes have more sugar; more important, they also have more extract, more acidity, and more flavor in the grape, which balances and carries the sugars. The categories of sweet wines, and their sweetness levels, are independent of residual sugar in the wine. Residual sugar levels are determined by the winemaker during fermentation, which is the process of transforming the natural sugar of the grapes into alcohol in the wine and carbon dioxide. So the dryness of a wine is independent of the ripeness level of the grapes upon harvest. If the fermentation is interrupted before all sugar is transformed, it will result in a sweeter wine. If the fermentation continues until little or no sugar is left, it results in a dry wine. Grapes for dessert wines have so much natural sugar that they will not ferment completely and residual sugar, or sweetness, will remain.

Now, residual sugars in wines that are not meant to be sweet happen when fermentation is too hot. Fermentation gets "stuck," which means that it stops before all the sugars have been eaten by the yeasts. This residual sugar can be glucose, fructose, or sucrose—the sucrose usually being the sort of sugar winemakers add if they chaptalize the wine. And this is usually the sugar not digested by the yeasts (which break it down into glucose and fructose). Back to ripeness: The degree of ripeness is always going to be a combination of sugars and acids. Simply put, the right amount of warmth, or heat, ripens the grapes, but too much heat makes them ripen too quickly and the rest of the plant cannot "catch up." This is what we mean when we differentiate between sugar ripeness and phenolic, or physiological, ripeness. The latter being when all the components of the plant ripen. It is a term bandied about a lot now. It never used to be, because it was never really a problem. In the Northern Hemisphere's cool climate regions, phenolic ripeness could be obtained through careful vineyard husbandry. In fact, underripeness was the issue, and it was usually a race to get the grapes ripe enough before any bad autumn weather arrived. And it was only in the Southern Hemisphere and warmer regions that the challenge was to get grapes to reach phenolic ripeness without making wines with huge alcohol levels and no natural acidity. There is no point picking earlier at 13 degrees potential alcohol in order to avoid an alcoholic wine, because if the phenolic ripeness isn't ready, the wine will have an unpleasant green, unripe flavor to it.

How things have changed. This tightrope between grape-only ripeness and whole ripeness is now walked by most. Extreme heat creates sugar ripeness only, and we are left with wines with "green tannins." Sugar ripeness just needs a lot of heat and sunshine, but to get a proper, all-over ripeness you need those long, cool hang times. The heat is pushing Mother Nature along, and she doesn't like to be rushed. Again, it is all about the balance among sugars, alcohol, acidity, and ripe tannins. Have too much or not enough of any of these components and the resulting wine will be unbalanced. I don't care what anyone says about their higher alcohol being "balanced" by their higher extracts or their phenolic ripeness. They are wrong. And to those that argue that fine wines can be made at over 14 percent, that alcohol is "simply a link in the chain of balance" and that you can find "alcoholic" wines at 12 percent if there isn't enough matter to balance that alcohol, and you can find superbly balanced wines at 15 percent—it should be pointed out that the crucial relevance is the degree of alcohol. Alcohol levels of 12 percent do not erase every other component of the wine, as do alcohol levels of 15 percent. There is a huge difference.

You can't have it all. If they got their phenolic ripeness (by picking late), then they unfortunately did so at the expense of the sugar and acidity balance, and got overripe grapes. And vice versa: If they picked early, in the hope of keeping the sugars down and preserving acidity, then they will not have achieved their phenolic ripeness and will have green tannins. They can't win for losing.

Investing in the Past

Assuming that by the valuation of that period their cost may be put at 100 sesterces per amphora, but that the interest on this sum has been adding up at 6 per cent. per annum, which is a legal and moderate rate, we have shown by a famous instance that in the principate of Gaius Caesar, son of Germanicus, 160 [AD 39] years after the consulship of Opimius, the wine cost that amount for one-twelfth of an amphora—this appears in our biography of the bard Pomponius Secundus and the banquet that he gave to the emperor mentioned: so large are the sums of money that are kept stored in our wine-cellars! Indeed there is nothing else which experiences a greater increase of value up to the twentieth year—or a greater fall in value afterwards, supposing that there is not a rise of price. Rarely indeed has it occurred hitherto and only in the case of some spendthrift's extravagance, for wine to fetch a thousand sesterces a cask. It is believed that the people of Vienne alone sell their wines flavoured with pitch, the varieties of which we have specified, for a higher price, though out of patriotism they only sell it among themselves; and this wine when drunk cold is believed to be cooler than all the other kinds.

—Pliny the Elder, *The Natural History*, Book 14, s. 55

CLIMATE WILL DESTROY THE MARKET IT CREATED

Since harvest reports are the key instruments upon which all projections concerning a particular's wine longevity are based, they are also the key instrument used to guide any investment in that projected longevity: the backbone of fine wine investment; that game that has warped the industry, and the wines, so unrecognizably. That game, so infused with irony. The traditional investment model is inane and climate change will be the final nail in its coffin. Future investment has to be focused on the future of the new energy technologies that will fuel our future wine regions, not on buying wine futures via the *en primeur* system. Even Thomas Jefferson thought as much when he wrote that buying wine directly from the château was the only way to buy it, stating that "it is from them alone that genuine wine is to be got and not from any wine merchant whatever" (Stephen Brook, *Bordeaux*, Mitchell Beazley, 2001).

The post–French Revolution bankers-*cum*-château-owners were considered "absentee landlords," very much like modern investors who are happy to remain far removed from their wine and leave it sitting in a bonded warehouse, untouched, unloved, and un-drunk (!), as they wait for it to increase in value. Wine was sold to *négotiants* in barriques who then bottled the wines for their customers, or exported them in barriques and the importers then bottled them. Some were very good at it, some were not. And the system was rife with fraud. The wines were diluted with water, wines from other regions or countries, and fruit juice. It was not until about 1920 that a few of the first growths started to bottle their own wines at the château. And it was not until as recently as the 1970s that it became the normal practice for serious producers.

The *en primeur* system gives the wine trade, or château, a two-year period during which the wines await retail release at higher prices (hopefully—this is what did not happen in 1970–71). Château owners in Bordeaux keep their wine in barrels, aging, for two years before bottling it. This is an expensive thing to do, a good thing, but it weighs heavily on the cash flow. So if one feels like putting on some rose-colored glasses one day, then the *en primeur* system could be looked upon as originally having been for the good of the wine and of the producer—a benevolent sort of investment. It is not at all certain that two years after harvest the wine will sell rapidly at the best price. The system counteracts this risk and provides a bit of insurance. But if the original intent was altruistic, it was quickly replaced by opportunism.

When a château offers its wine *en primeur*, one or several brokers de place are called upon to divvy up the wines among the merchants with whom the château has chosen to work. It is the broker who negotiates the wine's price with the merchant. In the past, when the system was ruled by the wine merchants, the brokers had an even more powerful role. But the economic crisis of the 1970s (which was then followed by the Cruse scandal) weakened the wine merchants of Bordeaux. Since then it has generally been the owners who have laid down the law, which only makes sense. They choose their wine merchants and the number of cases that will be allocated and at what price. So then, a "delicate arrangement" evolved. Merchants can turn down an owner's price and allocation amount, but they then run the risk of losing that château as a client.

KEEPING UP WITH THE LATOURS

I can certainly see that you know your wine, sir. Most of the guests who stay here wouldn't know the difference between Bordeaux and Claret.

—John Cleese (Basil Fawlty), *Fawlty Towers*

The owners usually set their wine prices by watching to see at which price their neighbor's wine is sold, accounting for the merchant's profit margin—a bit of Cabernet curtain-twitching. All the owners watch as the premier *tranches* are offered, the sort of "toe in the water" approach the owners use to test the market. The price of a particular growth is based on its reputation—no longer on its quality. Some of my colleagues would argue that it never was, and that the system has reinforced the perverseness of the classification system. In 1855, the rule of the day was "the most expensive wine must be the best one."

En primeur and its tangent topics merits a book unto itself, so here are just a few more interesting details about it. The owners do not sell their entire stock *en primeur*. There is no need to rush out and buy it all. But we are meant to believe that. Yes, in a great year, the greater wines will be more in demand, but it is not as though you will never have a chance again. In fact, it is only the consumer who ever loses out in the *en primeur* system. The investor/consumer is paying a merchant up front for wine that he or she will not see for eighteen months. And this payment is being called a "cash advance" by the *négotiant* for tax purposes, and neither title nor ownership of the wine passes to the investor/consumer until the wine is released by the château. The *négotiant* does not have possession of the wine, the merchant does not have the wine, and the consumer does not have ownership of it. And guess what happens when a merchant goes broke?

The system is a dinosaur and seems to be unwilling to acknowledge the changes unfolding before it. It is as though there are two worlds: the real and the virtual. In the real world, the world in the vineyards, everyone talks about climate change. They have to—we are standing there looking at it. But in the virtual world, the world of London investment firms and most wine industry trade journals, there is never a whisper. They are desperate to keep a lid on what may explode their game: Since 1988, when reliable data first became available, the fine wine investment market has generated an annualized return of 12.1 percent. Why mess with that? Fine wine consistently outperforms shares, bonds, and other asset classes, delivering annual double-digit growth. Fine wine investment is less volatile than other assets such as equities, gold, and oil while the correlation between financial and fine wine markets is relatively low, providing greater resilience to recessionary conditions. Fine wine investment is a tangible asset (if the buyer ever bothers to visit it in its storage facility); it is a finite and reducing supply as vintages are consumed versus increasing demand; it provides the ability to offset currency influences; and it is tax-efficient with its potential exemption from capital gains tax. And you can brag about it over dinner!

A recent wine reporter wrote, in his analysis of the poor 2013 Bordeaux "investment environment":

As it struggles to find direction, reports of the Bordeaux 2013 harvest do not provide much cheer. But does it spell further trouble for Bordeaux? Maybe not. (*Yes, of course it does.*) The market lives and dies on both demand and supply. Traditional collectors have not bought en primeur since 2009 (*And, have you asked yourself why that is?*) If they are still drinking, then their stocks are down. The small and mediocre 2013 vintage will be of little help. This might just be the bad news event that piques the market's interest—and subsequently leads to restocking.

This is so far off the mark, and such an irrelevant analysis, that it beggars belief.

So what does climate change have to do with this? Well, the system got further mired in the ambiguous "point" systems of American wine critics. Suddenly their opinions were having a greater influence on the prices than the guidance of the *négotiants*. All fingers pointed at Robert Parker, whose American palate had been trained with warm climate wine and who prefers the big, oaky, ultra-ripe wines. If he liked a wine, and gave it prizes and high scores, it sold. If he did not, it didn't. The Bordelaise started emulating this New World wine style, adopting the methods to achieve ultra-ripe grapes with the high-alcohol, sweet style that got the attention of the New World critics. As a wine judge myself, and having sat on hundreds of panels all over the world, I can vouch for the existence of a continental divide as far as palates go; it's an acknowledged given. And I can see how after a dozen series of a few dozen wines each, one's fatigued palate might be seduced by something that comes along and sweetly tricks you into thinking that it is a ripe, mature siren, when in reality it is an underaged, unbalanced juvenile wearing too much makeup and her mother's high heels. A properly trained professional is meant to detect the differences, and to reward them accordingly. That said, with climate change and winemaking styles, it is getting more and more difficult to distinguish grape varieties and wine age. But it wasn't, back in the 1980s, when the paradigm shifted.

A "NEW KIND" OF BALANCE—YOU MEAN AN "UNBALANCED" KIND OF BALANCED??

Higher temperatures cause more sugar, more sugar causes higher alcohol, higher alcohol, creates imbalance, and imbalance means a compromised longevity. "And then there's alcohol—and these days, lots of it. Everyone I spoke with agrees that the different chemistry of high-alcohol wines means that they will age differently from the old familiar 12.5%-in-a-good-year benchmark

Bordeaux wines of yesteryear . . . Throw in factors such as climate change (riper grapes) and changes in closures (screwcaps that alter the oxidation rate) and it's clear that the sensory profile for long-lived wines is a moving target" (Tim Patterson, "What Really Makes Wine Age Well?," *Wine & Vines*, July 2008).

> Neither Jacques nor I think you can make balanced wines at over 14% as we do everything we can to keep the levels down, including not cutting the yields back too much and picking earlier—sometimes two weeks earlier than the properties advised by Michel Rolland. The more you make the vines work, the more you turn them into Arnie Schwarzeneggers—blockbuster plants with highly concentrated grapes (Fiona Thienpont of Château Le Pin, Pomerol, as quoted by Andrew Jefford in *Alcohol Levels: The Balancing Act*).

And highly alcoholic wines are unbalanced, and unbalanced wines don't age well. These wines, these Bordeaux that are now unbalanced by alcohol—first by man's desire to please the critics, and now by climate change—have had their aging potential compromised and thus, their value as a commodity lessened. This commodity is losing its uniqueness and its reliability: the two basic tenets of investment. If these wines do not cellar as long as they used to, will this mean that investors will get a smaller return, but a faster one? Will the system remain in place but in abbreviated form? And will the investor/consumer accept the new Bordeaux profile? How sweet does it have to get before people say "no more"? If Médoc grows Carignan, how will that change the Bordeaux brand, and how will that go down? I asked a member of a well-known wine investment company this very question. We bandied about our theories, and his conclusion, ultimately, was that the wines are now such a "brand" that their quality no longer matters. The name is enough, and they will continue to get the asking prices. He may be right. But I hope not. Die-hard investors rarely taste the wines before they buy it. They just call up people like him and place an order. I call this "shopping by numbers."

A perfect example of this was found in a wine cellar I was asked to inspect while honeymooning on Capri. The hotel owner asked me to give him an idea of the value of the wines he'd inherited when he bought this villa built into the rocky south face of the island. It had been the home of the British singer Gracie Fields, and her brother-in-law had been in charge of stocking her famous cellar. After a bit of heart-quickening coasteering, we pried open the humid, crumbling wooden doors and peered inside. The smell was delicious. But the place was a mess. Behind the piles of Cliff Richard albums and a cricket bat collection, a sort of order was eventually revealed, clearly showing the remnants of this man's buying practices. He had methodically, and by the number, bought equal amounts of each of the red and white top growths from both Bordeaux and Burgundy, in every top vintage since the early 1900s. There was not one bottle that revealed either a personal preference or a hint of whimsy, passion, or knowledge. It was top-notch, textbook, blind purchasing to a t. Boring. Further, it had not been stored properly and half of them had their labels peeled off and their corks dried up. I came across a 1922 Meursault, the cork shriveled, the label unreadable, the bottle empty save for a few dusty, crystal remnants stuck to the inside's bottom. What could have been a collection worth hundreds of thousands was reduced to thousands, and worse, it had all been wasted and no one would ever enjoy those beautiful, beautiful nectars. That "textbook" list is going to change. It is being rewritten as we speak.

HOW WINE IS PUT TO WORK ON THE STREETS

Investment firm brochures all have pretty much the same content—and they perfectly illustrate the relationships among the wine critics, climatic taste preferences, and the warping of the wine industry by the investment sector.

Ever since fine wine was produced, it was used as a commodity. Wine investors find its inherently reducing supply (if they ever get around to drinking it!) irresistible, as it drives a frenzy of demand. This demand is primarily for the top thirty châteaux in Bordeaux and the balance of value sits with the five Premier Crus as well as the Super Seconds and the premier Right Bank wines. In my opinion, wine investment has warped the original intention of the *en primeur* system and it is now the wine investors who benefit from it, and certain critics who control it. The wine producers, the château owners, and the passionate, non-collecting consumers, are mere pawns.

THE ROLE OF THE CRITIC

One firm plainly and matter-of-factly boasts in its brochure: "The role of the professional critic within the wine industry is very significant. They provide an independent evaluation on the quality and potential value of wine and publish scores at key times within the industry calendar. The most preeminent critic with global influence is American Robert Parker Jr. Our wine selection policy is geared to Parker's scoring system and most fine wines offered to clients have a score of 95 points or more." Robert Parker hit the news again with his announcement that he would delay his 2013 scoring—which left the market in limbo and the campaign stalled. This was not the first time he did this. And it is a mystery to me how he is allowed to hold an entire market at ransom. If he delays his scoring, the campaign should go on regardless. Further, he has taken to rescoring his notes on a vintage ten years later. Surely this means that he is getting two bites at the apple? Regardless if whether he scores up or down, the market shifts and there are beneficiaries. This is not merely moving the goalposts, it is adding new ones!

THE RETURN?

Most firms deal in wine futures (*en primeur* wines)—buying wine at the earliest possible stage, "in barrel": before it is bottled. A firm will promote these investment possibilities (wines) in the spring following the previous harvest, and after the March barrel tastings performed by the critics. *En primeur* is considered to be a far riskier practice than simple wine investment. Firms provide a client with a Certificate of Allocation, which will be replaced by the Certificate of Ownership once the wine is bottled and shipped. This is because even though the client has paid for the wine, the ownership rests with the château.

What do investors get out of wine futures? As stated previously, tracking data since 1988 on all the trading Bordeaux châteaux tells us that with a typical wine investment portfolio, the five-year compound annual growth rate (CAGR) over the life of the index is worth the risk. The long-term average CAGR is 14.9 percent, with the average over the last twenty-four years at 12.5 percent, and 60 percent of the time, investors' returns have exceeded 10 percent, while 75 percent of the time returns have averaged more than 5 percent. But the more recent data also says that these returns have been decreasing for the past three years and have been hovering around 4 percent. This is what is fueling the fear over the eventual demise of the *en primeur* campaign.

INVEST IN THE FUTURE OF WINE

With the system waning . . . maybe it is time to remodel this system and change the focus of the investment? The classic Bordeaux model is already nonexistent . . . so what is there to invest in? Invest in the future of wine, I say. Invest in the land, the people, and the improvement of the technologies that we have today to make them more affordable, and the technologies that are going to be needed in the future. As the author Bjorn Lomborg, an adjunct professor and director at the Copenhagen Business School, states: "The smartest long-term solution—one supported by three Nobel laureates—is dramatically to increase funding for green energy research and development. The trick is not to subsidise today's hugely inefficient green technologies, but to focus on innovating down the cost of wind, solar and many other amazing possibilities. If future green technology were cheaper than fossil fuels, everyone would switch—not just well-meaning westerners." ("Climate Change Is Real, But We Have Time," *Sunday Times*, London, September 22, 2013).

Why not invest in the wine regions themselves? Look at what Hugh Johnson did for Hungary's Tokaji. Look at all the investment pouring into English wineries: Wineries in southern England are benefiting from temperatures that are suitable to champagne grapes, allowing good-quality wines to be produced. English sparkling wine is now retailing for similar prices as champagne: useful to know if you are a wine investor. Over the past forty years, the average temperature has risen by around 1.5° C, making a big difference to the harvests. The variability between wet and hot years is still there, but at least they now have ripe harvests again (England used to have vineyards in Yorkshire, during the Roman period). The best English sparkling wines have fabulous acidity and freshness. The dry whites are full of refreshing personality and character, and a few producers are playing around with Pinot Noir. In the next twenty years, English sparkling wines will only get better.

And what happens if eventually there is no viticulture in which to invest? Is that an eventuality? If so, wine investors, move over. After becoming used to investing and buying wines from new and different regions, what is next? Scientists fear that the removal of natural vegetation, chemical spraying, and processes associated with viticulture and the predicted changes in the world's wine map could have a devastating effect on native wildlife species, including the grizzly bear, pronghorn, and endangered giant panda. "Climate change will set up competition for land between agricultural and wildlife—wine grapes are but one example. This could have disastrous results for wildlife" (Hannah et al., *Climate Change, Wine, and Conservation*, PNAS). Lee Hannah says that conservationists and viticulturists will need to discuss the possible impacts of global warming before new wine-growing regions are opened up.

So if today's wine investors want to continue to see a return in their "product," their first concern will be protecting it for the future.

HARVEST REPORTS

Here is a selection of 2013 harvest reports from around the world. It makes for fascinating and entertaining reading. First, because it is amazing how many mistakes and misconceptions get spread in poorly written harvest reports, and that amuses me. Second, those that are well written—inevitably by the wine producers—are witty, engaging, and provide an intimate look into the lives of these people and show us just how close they are to their land.

Particularly enjoyable is the collection of harvest updates from the Tuscans. And the Georgian report is interesting as it emphasizes the fact that there was no "state intervention" this year: such an alien observation for other wine-growing regions.

Be careful when you source a harvest report. They are written by PR companies, appellation governmental bodies, wine sellers or agents, journalists, critics, and the producers themselves. Avoid those that are blatantly cheerful and upbeat about every year—they have an agenda! And there is a sort of language they all adopt. For example, when discussing yields, you often will hear: "Oh, despite the smaller yields, the grapes that did come in were healthy, small, and concentrated." Or, "Several heat waves resulted in low yields. On the other hand, the wines are already showing a high concentration." Well, yes, they would, wouldn't they? This is code for high sugar levels that translate into even higher alcohol levels.

Low yields are good. They are one of the greatest indicators of quality wine production. But *why* the yields are low is very important. Are they low because hail destroyed most of the buds in spring? Or because there was torrential rain during harvest (which leaves the remaining grapes diluted)? Or because heat and drought damaged the crops (which means the remaining grapes will be too high in sugar)? Or because the weather was perfect all year long and the producer practiced meticulous vineyard techniques and achieved a perfectly balanced result?

The 2013 harvest reports provide a good example of the current climate trends. Europe as a whole was divided between north and south, and then again, this same divide was seen within the different countries. Austria had poor weather during spring flowering for a smaller harvest, especially of Grüner Veltliner, while Greece lost over 20 percent of its crop to a windstorm and heat. In Italy, France, and Spain, the northern regions had extreme rain, and the southern parts, extreme heat.

So . . . here is a quick overview of 2013 as well as a small and random selection of full-length harvest notes from around the world.

You will be tested later.

2013 AT A GLANCE

ARGENTINA

A good season: a long, cool growing season, which meant that the grapes enjoyed a good long hang time before slowly reaching maturation, especially as the temperatures were up and down. This is good. Harvest stretched from early March to mid-April, with most winemakers reporting that they picked five to fourteen days later than usual. This is also good.

Spring rains and no spring frost meant more good news. The grapes came in with high acidity levels, low alcohol, and fine but solid tannins. Sounds very promising, especially for Malbec.

AUSTRALIA

South Australian regions such as Clare Valley, Barossa Valley, McLaren Vale, Eden Valley, and Limestone Coast experienced low yields caused by drought and heat. One of the earliest vintages in recent memory. Yields were down between 30 and 50 percent in the Barossa Valley because of hot weather and below-average rainfall, despite lots of irrigation to ensure that the grapes did not wither on the vines. The comparatively cooler regions such as Adelaide Hills and Coonawarra were able to escape the worst of the heat. But a heat wave meant that the harvest began very early, in mid-February. The growers frantically picked the grapes as their sugar levels mounted. The dry conditions meant that at least there was not much disease around. But the crop is small and overly concentrated. There will be some big, hot, and heavy wines produced. Winemakers have put on a brave face and unconvincingly pin their hopes on their Rieslings retaining a good acidity. Hmmm. If Austria is complaining of a lack of acidity in their Rieslings, what can that possibly mean for those that hail from here? After two previous cool and wet vintages, Victoria, Yarra Valley, Mornington Peninsula, Geelong, Heathcote, Grampians, and Strathbogie Ranges also had hot, dry conditions creating low yields. It was so hot that all the varieties ripened in quick succession and had to be picked at the same time—which creates a backlog in the winery. This was happening all over southern France, Italy, and Spain, too. Western Australia (Margaret River, Pemberton, Great Southern) fared rather well. Harvest weather was not as hot as in neighboring regions, but spring storms caused damage during *floraison* and reduced yields.

CANADA

A huge range of wine regions—too big to cover here. But here is what happened in British Columbia's Okanagan Valley. Their year was marked by warm, dry weather, which is unusual for them. Summer was nearly a perfect season, marked only by light rainfall in July. In fact, this year was the hottest many of the producers had ever seen, with a record in September for 91.2°F (32.9°C), rivaling their 1998 as the year with the most heat-degree days (remember Winkler) above 50°F (10°C) but below 86°F (30°C). For many of the world's regions, Okanagan's "hot" would be their "cool." Yields were above average, and the grapes were beautifully ripe and healthy.

CHILE

A nice, cool, slow growing season. One of the coolest vintages in the past decade: Highs were sometimes 5 to 8°C colder than usual. The only downside is that there was a bit of a struggle for some white varieties to ripen. But what a nice problem to have. And despite some scary frosts and hail reducing the yields, those grapes that did come in were undamaged. Harvests still started a few weeks later than normal, which means that the late-ripening reds such as Cabernet and Carménère should be well balanced as they enjoyed a long, cool hang time. We should see more elegant wines with lower alcohol levels and fresh acidity.

CHINA

It was difficult for me to get information on China: My Chinese is a bit rusty. Having never been there, and being unfamiliar with these new regions, my knowledge is truly lacking. But the information I did find was fascinating. Here is a small note on one region, the Shanxi region. "Chateau Rongzi is located in Xiangning. Main grape varieties are Cabernet Sauvignon, Merlot, Marselen and Chardonnay. Head of grape base said, flowering started on early June. And fruit set in the end of June. This year there were lots of rains, but no disease was found. And for grape base in Taigu region, flowering was delayed because of April's sudden big snow. Chardonnay started flowering from May 23 and ended on June 6. As temperature rose, Arlene grape first started *veraison* on July 12." (Original text in Chinese: www.winechina.com/html/2013/07/201307177361.html)

FRANCE

The year 2013 was one of the worst in almost thirty years—some say even forty. All Europe had an erratic and stormy year. But after the string of very hot vintages, something had to give. Cool temperatures, heavy rains and hail reduced yields, yes, but also damaged the quality of the grapes that did come in. The harvest should total about 43.5 million hectoliters. The ten-year average is 45.5 million. In Burgundy, devastating hailstorms in July caused enormous damage to their vines and destroyed their crops. A few weeks later, similar storms ravaged Bordeaux with huge nuggets of hail. But after the cold, wet spring devastation, the summer was warm and dry. The grapes are showing low sugar content, which could be a good thing. If the weather didn't dilute the wines' acidity and fruit, the year may turn out some classic surprises.

NEW ZEALAND

Mother Nature worked her magic on the North Island's yields, lowering them with spring frosts, but the growing season was long, with a good summer, with harvest beginning on March 15. The fruit that did come in was clean, ripe, and healthy with plenty of acidity. This bodes well for balanced, fresh wines. I look forward to trying the Pinot Noirs. The South Island regions (Marlborough, Canterbury, Waipara, Central Otago, Nelson) had a similar season, although it sounds as though they experienced more of an extreme difference between their seasons. They went from spring frosts, to a summer warm enough to mature the varieties at the same time and to cause an early harvest. Apparently, everyone was running around trying to get everything in at the same time before they got any riper, and earlier than usual. Those producers who didn't move fast enough got caught out by a late heavy rain—which will mean dilution.

PORTUGAL

Like the rest of southern France, Italy, and Spain, Portugal's Duoro Valley and Setubal regions had been having a scorching growing season, the hottest it had known in eighty years, with a recorded midsummer high of 120°F (49°C). Then, in the first week of October, huge rainstorms hit, heavier than had been seen in years. A year of dramatic extremes. Some grape varieties had already been harvested and were safe, if not a little overripe, but the varieties such as Moscatel and the Touriga Nacional will risk dilution. But with such high alcohol, maybe some dilution will end up balancing things out? Dry wines will suffer, but port should be great.

SOUTH AFRICA

A very challenging vintage with erratic weather: good winter and spring, but then rain and humidity midharvest. Those that cut back the canopy to expose the fruit and keep it healthy and those that were very careful with their grape selection at triage will be the winners. In the warmer Swartland region where the Rhône varieties are favored, there was quite a lot of heat, which stressed the vines. But they were then rescued by some light rain in the end. In general, an erratic vintage with good yields, as the majority of the grapes enjoyed cool, long hang times and should produce some balanced and elegant reds.

UNITED STATES

Winemakers across the board harvested as much as two to three weeks early. Many Sonoma Pinot Noirs were in before mid-September. Warm weather, even in the coastal regions, brought Cabernet Sauvignon and Merlot in ten days earlier than last year. Producers in Washington State experienced the second hottest summer on record and also had to harvest early. The yields were up and the heat also means that the fruit was small and concentrated, so expect big, alcoholic wines this year.

The History Found in Harvest Reports

Red Bordeaux Harvest Notes 1771–1979: Excerpts from Michael Broadbent's *The Great Vintage Wine Book*

This is a fascinating slice of Bordeaux's climatic history. There is no better way to understand how the vagaries of weather can form a wine's character and aging potential. A bad year? See 1861 or 1951. An ideal harvest? See 1870, 1900, 1929, or 1953. Note that a "great" year was one in which the wines exhibited so much tannin that they were undrinkable right away. In fact, it is crucial to understand that Mr. Broadbent includes his tasting notes of these famous vintages from tastings he attended in the 1960s and 1970s. So each entry has a climate/harvest report written from that period concerning the year in question, but the tasting descriptions of the vintages are written *decades* later. That is *truly* remarkable and it is sadly no longer the case. But that is the point of the exercise—great weather makes great wines and great wines are meant to last a long time.

Note also the descriptions of what is deemed a good wine and not. It does not seem as though the English palate has changed its preferences too much, does it? Most of these notes would be perfectly at home in any of our current wine magazines. And I love how the lighter, less complex, "flavourly" wines are immediately shipped off to the Americans and Germans! Clearly, our palates have evolved. Very telling is his use of the descriptor *hot-vintage character* in his tasting notes and his inclusion of the following definition in his glossary: "The smell or taste of grapes baked in a particularly hot summer sun or from an area with a normally hot climate tends to be associated with high alcohol content."

This collection of harvest notes so directly reflects the events of our times. Of the dismal decade of the 1930s, he writes, "It seems that bad weather, bad wine and bad times go together." And, touchingly, the text is peppered with references to the devastation of the world wars: "average crop partly due to labour shortage." It is also interesting to follow the trail of the dreaded phylloxera beast through his notes. Michael included harvest and tasting notes for every single vintage. Here I've collected just a sampling. I implore you to try to obtain a copy of this tome. Published in 1980, it still makes for a riveting read—a veritable roller-coaster ride. And it is highly relevant—even more so today, I daresay.

1784—"It is one of the best vintages which has happened in nine years . . . Chateau Margau [sic] bought by myself on the spot." (Thomas Jefferson)

1798—A great year: a good crop of excellent, full-bodied, velvety wines." (Christie's catalog)

1799—Picking began on October 5; an above-average crop.

1803—Picking began 25 September 25. A good year.

1811—Famous year. Picking began September 14. A magnificent and abundant vintage.

1815—The year of Waterloo. "A remarkable year in every respect. The wines agreeable and perfumed, having both body and richness."

1816—Very wet spring and summer; disastrous.

1823—Rain during harvest.

1825—Picking began September 11. A celebrated vintage.

1828—Warm summer, rain during vintage; light but fine perfumed claret.

1829—Cold and wet; green, unripe grapes.

1840—Abundant; good color and flavor but "its want of body unsuitable for the English Market."

1844—Picking began October 7. A first-rate vintage, best since 1815. Below-average crop (yield).

1847—An abundance of light flavory wines not to the English taste of the time. Practically all shipped to Germany, Holland, and the United States.

1858—Favorable winter and spring, hot summer. A good crop; picking began on September 20. Wines have deep color, power, sève, fragrance, and finesse. Note that this vintage would be the first of two decades that would become known as the Golden Age of Bordeaux. For a cool region accustomed to struggle for ripeness and maturity—things were heating up.

1864—A truly *grande année*. Great heat prior to picking on September 17. Wines of pronounced bouquet, sap, finesse, softness, and elegance.

1865—Another magnificent year. Favorable conditions led to the early picking (September 6) of perfectly ripe grapes.

1868—Curious conditions: hot May to July. Exceptionally wet in August, extremely hot before and during the early vintage, September 7.

1870—Vying with the 1864 as the greatest pre-phylloxera vintage, possibly of all time. Frost reduced the crop, July heat-baked the grapes, and the vintage started early, on September 10, in excellent conditions. The wine was massive, *très corsé, très vineux*, but so hard and unyielding it took fifty years for it to be drinkable.

1875—An even greater crop than 1874, in fact the biggest production of wine in the Gironde from the start of records until the mid-1960s. This due to favorable weather, but also greatly increased acreage under vines following nearly twenty years of prosperity. But the wines lacked body. They were so lacking in tannin that most were drunk within two years.

1877—Long wet winter and spring. Phylloxera "making rapid progress." Picking began September 26.

1878—Described as the last of the pre-phylloxera vintages. In fact, phylloxera started in the Bordeaux vineyards in the mid-1870s, and wines continued to be made from ungrafted vines until after the turn of the century. What is certain is that 1878 was the last of a string of quite exceptional vintages. A growing year of variable conditions but favorable from September 1. Picking began September 19. Big crop, very good wines.

1881—Difficult climatic conditions. Phylloxera spreading to the Médoc; Château Lafite affected. Smallish crop of hard, charmless wines.

1885—Picking began September 29. Half crop due to mildew and the ravages of phylloxera. Ordinary wines.

1887—The best vintage between 1878 and 1893. A warm summer and excellent harvesting conditions from September 19. Still only half the average crop though mildew under control. Sound, healthy, full-bodied wines, now fading.

1892—Picking began September 22. Half-sized crop following two severe frosts and a 110°F (43°C) shriveling sirocco mid-August. A hailstorm then ravaged Pauillac. Irregular quantity and quality, little color but elegant.

1893—A complete contrast to the previous fifteen rather dismal years: no ravaging by frosts, diseases, or pests. Warm spring, early flowering, a baking hot summer and the earliest start to the harvest on record, August 15. Biggest crop since 1875, good quality.

1895—Very uneven conditions. Wet until end July; good weather August and September—but devoid of any beneficial rain. Picking began in exceptional heat, which made winemaking more difficult. Some grapes raisin-like.

1899—Abundant harvest. *Très grande année:* outstanding.

1900—One of the most perfect vintages ever. A superabundant crop of very fine wines. Hotter than 1899. Excellent weather conditions through to harvest September 24.

1906—Singularly favorable weather conditions but excessive heat and drought in August reduced yield. Good harvest September 17. Robust wines of high quality.

1914—Fine May, then an extended and delayed flowering. Hot August. Harvest September 20 in good conditions. Average crop. Turned out to be disappointing.

1917—Good, light, supple, fragrant wines. Warm May, good flowering; excessively hot June; July and August poor, changeable. Moderately early vintage, September 19, in good conditions. Average crop partly due to labor shortage.

1920—The first unqualified *grande année* after 1900. Wet, mild winter; excellent spring, growth advanced, no frosts; flowering perfect then July and August exceptionally cold. September fine, with rain to swell the grapes, but yield at harvest was one-third of 1919. Turned out to be a great vintage—held on.

1921—A year, like 1895, noted for its exceptional heat and attendant problems of vinification. April frosts, hail, excellent flowering then exceptional and continuous heat. Picking commenced September 15. Moderate quantity. Baked grapes have thick skins and high sugar content, resulting in deep-colored wines full of alcohol and tannin.

1926—*Une très bonne année* but small crop due to poor flowering after long winter, cold spring, and drought during long, very hot, dry summer (similar to 1865), which continued through to late harvest, October 4.

1928—A monumental year: flowering under good conditions; splendid summer, excessive August heat tempered by beneficial rain; harvest September 25 in promising conditions. Of all the vintages of this decade, the most massively constituted and holding best.

1929—Considered the best vintage since 1900. A complete contrast in style with 1928. Soft, elegant wines of great finesse and delicacy. Flowering started in good weather but became irregular due to rain. July, August, and September were unusually hot and dry, though there was beneficial rain mid-September prior to picking on 26th. The best (and best-kept) wines still retain the hallmark of finesse and suppleness but, less sturdily constituted than '28s, they are fading; those left should be drunk.

1934—The best vintage of the 1930s. Drought conditions in June and July were followed by sufficient rain to swell the grapes for a good, early (September 15) and abundant harvest: double that of 1933. Sound and deep wines, the best still excellent. But they are not improving and some are distinctly cracking up.

1937—A curious year: drought without excessive heat from May to the second week of September, when there was welcome rain. Harvest September 20, under ideal conditions. Quantity average, quality thought promising at the time but the wines, while often fragrant, are mostly dried up and astringent.

1945—One of the greatest of all vintages. The crop was severely reduced by heavy May frosts followed by hail, disease, and exceptional drought during the summer and during the early harvest (September 13). By and large they are great wines: deep, concentrated, packed with flavor. Some, however, are now drying out.

1947—Another postwar milestone. Good late spring, fine June; successful flowering; fine July; August hot with some night rain; fine September. The picking beginning on 19th in almost tropical heat. The trade press reported that the sugar content had reached unprecedented heights, which is why some wines were "pricked" (acetic smell, tart), and many show signs of acidity taking over. But there are some rich, ripe, exciting wines.

1948—Weather conditions, the cause of character, were somewhat perverse. Exceptional heat in mid-May—110°F (43°C).

1949—A great vintage. Fine, supple, beautifully balanced wines. Extraordinary weather conditions. January and February the driest months on record; flowering in cold and rain caused most disastrous *coulure* ever remembered; second half of June fine and warm; July the hottest and driest ever, equaling 1893. Temperatures of 145°F (63°C) recorded in the Médoc on July 11. [As hyperbolic as this figure might seem, it is confirmed by multiple references by growers in the Médoc at the time]. August lacked rain. Storms early September; harvest began on 27th in fine, hot weather tempered with a little benign rain. October driest on record. Quantity below average, quality *très bonne année*.

1953—A personification of claret.

1959—The vintage of the century. One of the most massively constituted wines of the postwar era. (It has its detractors; "too little acidity," "cracking up.") One of the finest Februarys in living memory—frost at night, early-morning mist, and hot sun; March equally good; April hot, cold, rain, and storms; May improved later; June good; July almost too hot; August fine and warm; September hot; a lot of rain from 13th but good weather for picking, which started on 23rd.

1961—One of the four best of the century. Deep, rich, concentrated, long-lasting. Nature effected a severe pruning; the remaining grapes, small and concentrated due to drought, benefited from the abundance of nutrients (drawn from the soil), which would otherwise have been distributed among a normal-sized crop.

1966—An excellent vintage. Stylish, elegant, well balanced. Lean rather than plump, though with good firm flesh. The lack of sunshine in August was made up for by a very hot sunny September.

1967—Quite attractive when young due to hot and dry August. They now resemble a girl well past her bloom of youth, whose heavy makeup is wearing thin.

1970—An outstanding vintage with that rare combination of bumper crop and high quality. Great heat and drought in July; August rainy and cooler with hot intervals; beginning September cold and stormy but then a long run of sunshine during the vintage.

1973—A fairly light and abundant vintage. July wet with little sunshine; August exceptionally hot and sunny; end of September rain. A huge crop. Needs drinking fairly quickly.

3

What Is Being Done: Mitigation and Adaptation

We can see what is going on in the vineyards and we can taste the effects. Though all producers will be experiencing the changes in their own way, is it possible to adapt to these changes and, eventually, to retard them? The problem is that most adaptation techniques lessen the quality of the product. For example, planting vines in shallow soil to reduce their water consumption, or introducing irrigation, or increasing the canopy's vegetation to protect the grapes from the sun, are all practices that will produce a lesser wine. Also, there will be a point at which a region cannot adapt any more: The law of diminishing returns applies. After adaptation comes cessation . . . you pull up roots and move. There will be no choice. Wine will have been diluted to pointlessness. In certain regions, adaptation will have been taken to such an extent that wine will no longer possess its unique identity. Further, the principal adaptation technique, irrigation, is also number one on the most-wanted list of mitigation criminals. Mitigation and adaptation are hurtling toward each other and will eventually collide and negate each other. More serious still, forced migration as an adaptation technique will pit *Vitis vinifera* against other crops and members of the animal kingdom in the competition for food and water.

The best way to increase a supply of anything is to decrease the demand for it. Stop using as much. Basta. Perhaps, as is often the case, lessons from the Past become our Future. It wasn't very long ago that the only choice we had was to drink the wine that was grown down the road from us. Yes, international wine trade is hardly new, but our vessels are no longer horse-drawn nor windblown. Spanish wine educator Pancho Campo, a knowledge-able climate change activist who organized the World Conference on Climate

Change and Wine, reminds us that "the solution is not just in changes like new varietals or replanting vineyards. The solution is mitigation. It's more important to reduce CO_2 emissions and energy waste and increase recycling at the winery level."

A queer little quirk in these proceedings, a potential silver lining, is that during this adapting and reshuffling period, there will be more regions making wine than ever before: The regions that are benefiting from warmer temperatures will be producing more of their own wine, more reliably; along with those regions that will commence viticulture entirely; as well as those regions that are going to continue struggling to produce before their demise. This then should mean that there would be a reduced demand for wine imports, assuming that everyone drinks their own production. Wine's biggest cost to the environment, apart from water, is transport. Figures in 2007 already confirmed the UK as the world's largest importer, bringing in over a billion liters a year: 50 percent from the New World and 80 percent in bottles weighing an average of 500g each by ship and trucks. The wine industry accounts for a quarter of the 1.5 percent of the 670 million tonnes of the UK's annual greenhouse gas emissions made by the alcohol industry. Transport emissions are highest during road travel, then sea, with rail travel having the lowest levels (WRAP, GlassRite Wine, May 2009).

Imagine, then, if every country just drank whatever alcoholic beverage it was able to produce, whether wine or not. It may come back to that. And would it be the end of the world? I could very happily survive on a diet of margaritas. Wine is, and should be, considered a food staple, a nutritious part of our daily diet—therein lies the natural beauty and effortless majesty of it. But the artistry of wine was stripped away when it was allowed to become a valuable trade commodity. Then it became a luxury item, and we became such spoiled, all-consuming, brand-hungry brats. There will be a point at which we all feel rather frivolous, worrying about whether our wine at dinner is a local brew or has been flown in from Tuscany. When it comes down to having to decide whether to water our wheat crops or our wine grapes, we shall feel very silly indeed.

THE ENVIRONMENTAL COST
OF VITICULTURE

When we think of vineyards nestled in soft, lush hillsides, the images conjured are of an inoffensive, passive nature. What could be gentler on the landscape than a bit of fruit farming and fermentation? So Pieter Bruegel and *The Harvesters*. Ah, but since the 1980s, the world's thirst for this fermented fruit juice had knocked the wine industry up a few notches and transformed it into a mass-market giant, with its accompanying noxious side effects. No longer are the small artisanal domains, with their few acres to tend, the majority members of the club. Both in Europe and the New World, huge, industrial operations, some with planted acreages the size of an entire French appellation, churn out the stuff of supermarket dreams. The huge amount of land used, as well as the amounts of synthetic fungicides (sulfur is approved for use as a fungicide in organic wineries, but the highly toxic runoff from irrigation and rain is fouling rivers and waterways), herbicides, fertilizers, nondegradable materials, and harmful fuels—not to mention the processing (water usage), manual labor forces, packaging (including bottle weight), and transportation (huge carbon imprint)—all have a much larger impact on the environment than we imagine. The facts don't seem to fit the image, but they are there, front and center.

WHAT TO DO ABOUT IT?

Gregory Jones of the University of Southern Oregon, the leading researcher in this field, states that "although uncertainty exists in the exact rate and magnitude of climate change in the future, it would be advantageous for the wine industry to be proactive in assessing the impacts, invest in appropriate plant breeding and genetic research, be ready to adopt suitable adaptation strategies, be willing to alter varieties and management practices or controls, or mitigate wine-quality differences by developing new technologies."

The relationship between wine quality and temperature is well-established among wine producers. Cool climate wine regions have always produced the acknowledged finer wines. This relationship is unlikely to change, and the recent warming in the western United States is testing it significantly. The ability to adapt to the increasing temperatures will rely heavily upon a region's location and its legislative infrastructure. The *Vitis vinifera* is diverse and adaptable, but there are few varieties that can continue to produce quality wine in regions that have excessive heat. Wineries that cannot change locations due to lack of space or a strict appellation system will have to try heat-resistant clonal experimentation, screening, canopy management, new grape varieties, and, unfortunately, increased or new irrigation practices. Those that can will have to move higher in altitude or to coastal regions (for now). The problem is that the climate may change faster than the adaptation practices can be tried and tested. "The rate of climate change and/or the rate at which variations in environmental tolerance can be exploited may therefore impose adaptability limits, particularly in long-lived systems such as premium wine, for which the long time to maturity (1–2 decades) and in-place lifetime (3–5 decades or more) increase the investment and opportunity costs of changes in location or variety, as well as the potential loss should the actual climate change be different than anticipated" (Diffenbaugh et al., *Climate Adaptation Wedges*).

I think that the New World wine regions will at first have the initial adaptation advantage, as they are free from appellation laws. They are still in the process of experimenting and refining their plantings, so these climatic changes will be throwing them off piste, but they are already armed for change, so to speak. There are already new plantings in Patagonia and Tasmania . . . the movement to cooler climes in the south is already afoot in the Southern Hemisphere. Meanwhile, the Northern Hemisphere will be mired in bureaucratic red tape and a fearful clinging to the safe "knowns" of the past. But once this phase passes, it will actually be the Northern Hemisphere that will have the advantage, as it has more land northward in which to expand. But this won't happen until planting and irrigation laws are further loosened. Whereas the Old World had just settled in and thought that they were finished with all that, and now they will have to start all over again. And if you own only 6.5 hectares in the Côte de Nuits (thank you, Napoleon) and produce thirty thousand bottles of wine a year, what do you do? Where can you go?

On the opposite side of the coin is mitigation. Does mitigation have to involve technological intervention? Inge Kotze, the coordinator of the World Wildlife Fund's Biodiversity and Wine Initiative in South Africa, thinks that the focus should be less on carbon reduction through technological intervention than on better management of the natural systems Nature already has in place. As an example, she cited the case in which ecosystems "spring back into balance after the simple act of removing invasive plant species like pine trees or blue gum, which can suck a watershed dry." She adds, "Streams which farm workers who have lived in the area for 60 years have never seen flow, now flow" (Iverson, *Time*, 2011).

A lot to think about.

LEADING THE MITIGATION MOVEMENT

"Organic" . . . "Green" . . . "Biodynamic" . . . "Sustainable" . . . "Carbon neutral"—all these confusing terms are freely bandied about these days. What do they mean and how do they interact? *Green wine* is the sort of main heading under which the other terms are subsets. Organic wines have been around for a long time, since long before they became a marketing trend in the New World. In fact, many wine producers, especially small European domains, don't seek to be officially labeled as "organic" and jump through all the ensuing administrative hoops and endure the ensuing scrutiny, because, they argue, the practices used in organic winemaking are the sort of things that they have always been doing anyway. Organic winemaking, before it got its name as such, was simply a body of conservative, respectful, and non-interventionalist winemaking practices. It was only the segment of the wine industry that went overboard in technology that, when deciding to rein back in, made a huge deal about giving it a name, sticking this on their bottle labels, and turned it into a marketing angle. Biodynamics have also been around for a long time. This practice follows the moon and tidal cycles to determine vineyard work and harvesting schedules. The aeration and balance of the soil are also important. Nicholas Joly, the "nature assistant" who assists nature in creating his Loire Valley wines, has to be considered the modern father of the practice, and his books on the matter are required reading: *Wine from Sky to Earth* and *What Is Biodynamic Wine?* Whatever the consensus on these practices, it is difficult to argue the fact that if we take care of the soil, of the earth, then the fruit born from it will also be improved.

It should also be noted that organic and biodynamic wine practices can have a direct effect on the taste of a wine—hopefully, a beneficial one—whereas the practices adopted by a carbon-neutral program are more directly related to reducing energy and less about improving the quality of the wine. That said, mitigation practices such as reduced tilling, and reduced or non-irrigation, do eventually have a favorable impact.

TREADING LIGHTLY: CARBON NEUTRALITY

The new breed of green wine is carbon-neutral, and I have even heard of wineries taking this a step farther and going "natural" carbon-neutral. Carbon neutrality means leaving a net zero carbon emissions footprint, by either balancing a measured amount of carbon released with the equivalent amount offset, or by buying enough carbon credits to make up the difference in released carbon. Sounds good, but how effective will any attempts at carbon neutrality be if they include the use of carbon offsets in the equation? It makes no sense to allow heavy polluters to buy the offsets of light polluters when the world is an enclosed system and the effects of pollution will be felt by all. This is just passing the problem along as opposed to solving it. The Rolling Stones and Pink Floyd may have held carbon-neutral tours, but if they flew everywhere in jets, then to whom did their carbon use get transferred?

To be considered carbon-neutral, there must be a carbon footprint reduction to zero. But what to include in the carbon footprint depends on the company and the standards they are following: There is no set, internationally agreed-upon method or standard. In fact, it is all very confusing, as there are so many different entities acting as qualifying agents. But they all seem to be working toward the same goals and criteria. In general, it is accepted that there are two sorts of emissions: direct and indirect. Direct emissions include all pollution from manufacturing, company-owned vehicles, reimbursed travel, livestock—all sources that are directly controlled by the owner. These must be reduced and offset completely. Indirect emissions are all emissions from the use or the purchase of a product: A power company's greenhouse gases are a direct emission; the office or company that purchases that energy considers it an indirect emission (Wikipedia). They can be reduced with renewable energy purchases—which brings us back to the beginning, doesn't it?

Perversely, a perfect zero carbon neutrality cannot be achieved by wineries, as the winemaking process inherently emits carbon dioxide during fermentation. Still, it has to be a step in the right direction, and the number of wineries around the globe that have pledged themselves to carbon neutrality is inspiring. In the context of a winery, this means calculating the winery's total damaging emissions, reducing or mitigating them however possible, and then dealing with the remaining footprint by investing in carbon credits.

Campo Viejo in Rioja, for example, has reduced the weight of its bottles from 550 to 380 grams and is using electricity sourced from renewable energy. It's also participating in a project backed by the UN and validated to the Verified Carbon Standard (VCS), which is the building of a run-of-river hydro project on the River Heihe in Gansu Province, China, thereby offsetting their remaining carbon emissions. The Boisset family led the biodynamic move with their Burgundy estates, and in California plan to be 100 percent solar-powered this year. Blau Nou in Spain has been certified Carbon Zero by ZeroCO2 and uses sustainable farming techniques and hand-harvesting. Grove Mill Winery in New Zealand declared itself the world's first carbon-neutral winery in 2006. The winery reduced greenhouse gas emissions through the CarboNZero certification program, which has helped New Zealand wineries lead the world in offsetting their carbon footprints.

Parducci Wine Cellars in Mendocino, California, was declared carbon-neutral by California's Climate Action Registry back in 2007.

Cullen Wines in Margaret River, Australia, make 100 percent of its air travel and fuel-usage emissions carbon-neutral. Cono Sur Vineyards in Chile became the first winery in the world to obtain carbon-neutral status in terms of delivery, in 2007. Backsberg winery in South Africa won the Mail & Guardian Greening the Future Award for Energy Efficiency and Carbon

Management back in 2007, and was one of the world's first wineries to achieve carbon-neutral status. Tinhorn Creek Vineyards in British Columbia became the first Canadian winery to complete the Climate Smart program and offset its carbon footprint. Wakefield Wines in Clare Valley, Australia, has been declared 100 percent carbon neutral, based on a Life Cycle Assessment Model. The Australian winery considered everything in the wines' life cycle, from harvesting of grapes to recycling the wine packaging (www.carbonneutralkawarthas.ca).

Entire countries are in on the act. The Wine Council of Ontario has developed "best-practice" guidelines, in a program called Sustainable Winemaking Ontario, that cover aspects from viticulture and wastewater management on the farm to efficient energy usage in all winery operations. The intent is that this "from soil to shelf" program will help sustain Ontario's success on the world stage and is based on the belief that large-scale use of synthetic chemicals, such as fertilizers, pesticides, and herbicides, prevents grapes from offering their truest expression and disrupts the composition of the soil.

Can all this have a true mitigating impact? Are these forest-planting enthusiasts barking up the wrong tree? Which leads to another point: We shouldn't be rushing into planting trees wherever we can, especially not in areas where nature has already indicated that they are not meant to be, such as the northern tundras. Again, meddling with nature can have its repercussions, as Ken Caldeira at Stanford's Department of Global Energy reminds us: "Trees [planted] in snowy regions absorb a lot of sunlight that otherwise would have been reflected to space by the snowy fields," thus creating more warming. There is neither an easy answer nor one blanket trick to fix it all. But winemakers around the world are experimenting and exploring as many options as they can. "We're basically talking self-preservation here. I mean really, if it's not in the interest of wine producers to have stable, predictable weather patterns for growing specific grapes with delicate character, then we should all just hang it up and start drinking beer instead," argues Michael Stewart of Carbonfund.org.

I think it is overly ambitious to assume that such measures can reestablish "predictable weather." That ship has sailed. It cannot do any harm to clean up our act, however. Still, the bottom line is, we can placate ourselves with our inventive carbon offsets projects, our low-wastage systems, our new energy sources and tree-planting, but all these measures are part of the problem. Creating new technologies creates new pollutions. The real issue is overconsumption, overdemand, and overuse.

For a full list of the world's carbon-neutral wineries, visit www.theclimate changeandwineportal.com.

What a Winery Can Do . . .

- Use a carbon merchant for tree-planting (a fee is paid to plant the number of trees needed to take out of the air the same amount of CO_2 as is emitted by the winery).
- Use less glass in bottles to reduce weight and shipping costs.
- Reduce use of herbicides and fertilizers.
- Abandon glass and use plastic PET bottles or Tetrapak.
- Use photovoltaic and solar panels to generate energy and hot water.
- Introduce wind farms.
- Complete energy-efficiency audits.
- Refit winery and warehouse lighting, replacing incandescent bulbs with fluorescent.
- Replace old machinery with more energy-efficient models.
- Increase tank insulation.
- Create and use biodiesel or ethanol in vehicles and farm equipment.
- Turn animal wastes in methane gas, with a methane digester, for heating.
- Water management in general.
- Store and use rainwater for irrigation.
- Ship wine in bulk and have it bottled at the destination; reuse bottles.
- Hand-harvest rather than machine-harvest.
- Ensure new buildings are carbon-neutral, and retrofit where possible.
- Employ gravity-fed operations.
- Engage in organic and biodynamic farming.
- Increase planting density—better land use.
- Reduce tillage and machine-harvesting.
- Soil stabilization.
- Canopy management and pruning.
- Trellis modification to ensure aeration.

GLOBAL CHANGE IN AREAS SUITABLE FOR GROWING WINE GRAPES THROUGH 2050

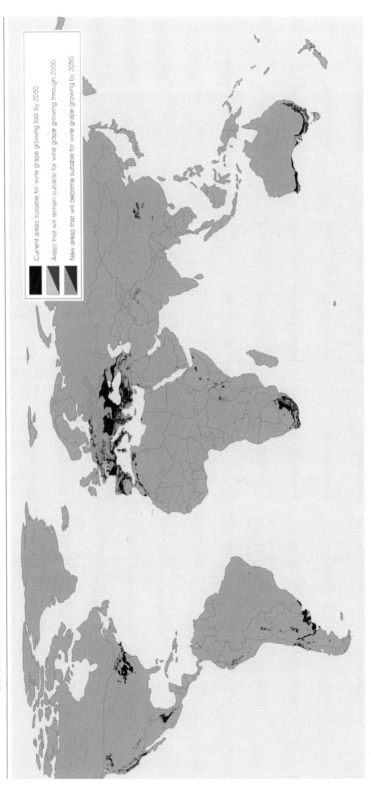

Current areas suitable for wine grape growing lost by 2050

Areas that will remain suitable for wine grape growing through 2050

New areas that will become suitable for wine grape growing by 2050

CONSERVATION INTERNATIONAL

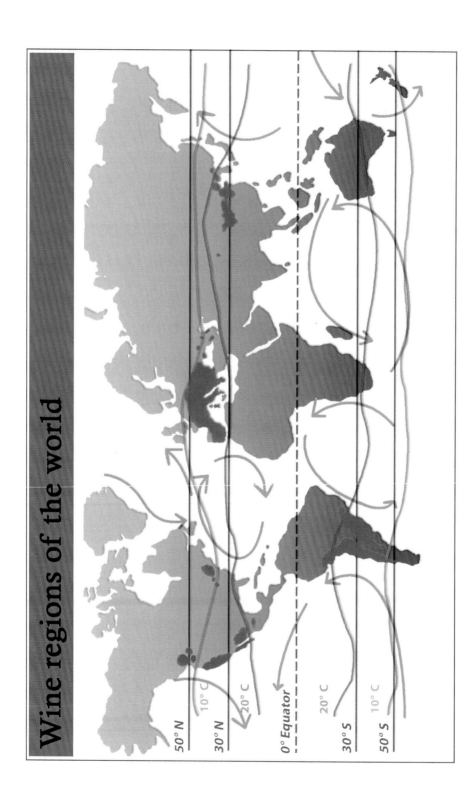

Wine regions of the world

CLIMATE CHANGE MAP OF EUROPE

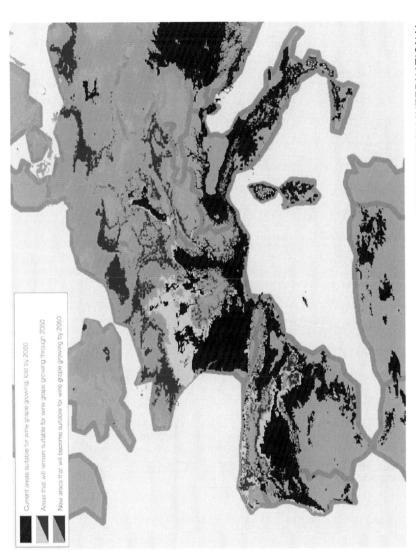

Current areas suitable for wine grape growing, lost by 2050

Areas that will remain suitable for wine grape growing through 2050

New areas that will become suitable for wine grape growing by 2050

CONSERVATION INTERNATIONAL

World Map of Köppen–Geiger Climate Classification

projected using IPCC B2 Tyndall SC 2.03 temperature and precipitation scenarios, period 2001 to 2025

Main climates	Precipitation	Temperature
A: equatorial	W: desert	h: hot arid	F: polar frost
B: arid	S: steppe	k: cold arid	T: polar tundra
C: warm temperate	f: fully humid	a: hot summer	
D: snow	s: summer dry	b: warm summer	
E: polar	w: winter dry	c: cool summer	
	m: monsoonal	d: extremely continental	

Af Am As Aw BWk BWh BSk BSh Cfa Cfb Cfc Csa Csb Csc Cwa

Cwb Cwc Dfa Dfb Dfc Dfd Dsa Dsb Dsc Dsd Dwa Dwb Dwc Dwd EF ET

Resolution: 0.5 deg lon/lat

Version of May 2010

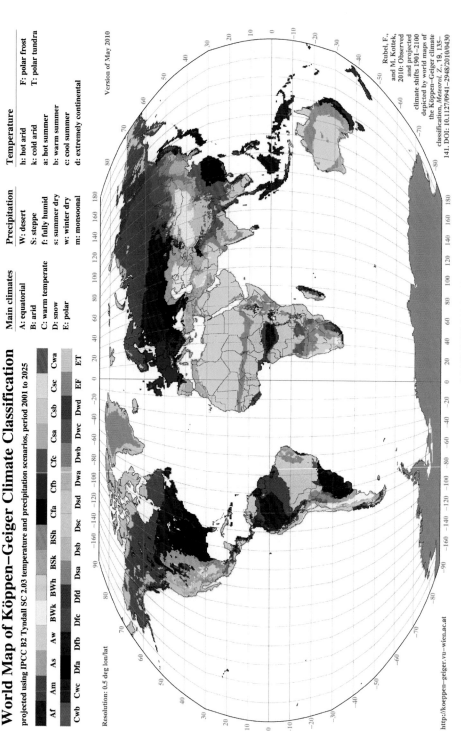

Rubel, F., and M. Kottek, 2010: Observed and projected climate shifts 1901–2100 depicted by world maps of the Köppen–Geiger climate classification, *Meteorol. Z.*, 19, 135–141. DOI: 10.1127/0941–2948/2010/0430

http://koeppen-geiger.vu-wien.ac.at

CARINTHIAN INSTITUTE FOR CLIMATE PROTECTION (KIKS)

- Use screening for shade (expensive).
- Use spot-spraying and pheromones for pest mating disruption.
- In the Northern Hemisphere, adjust slope exposition to northern-facing and not southern.
- Change or ignore local appellation laws.
- Plant warmer climate grape varieties.
- Reduce irrigation (to mitigate).
- Increase irrigation (to adapt).
- Select clones that withstand temperature and drought.

ADAPTING TO THE HIGHER ALCOHOL LEVELS

The problem that most immediately has to be addressed and solved, where possible, is the rising alcohol levels. As mentioned previously, climate change is not the only factor in creating higher alcohol levels; we also have to consider the influence of the modern and improved viticultural techniques as well as the stylistic preferences chosen by those wine producers who wish to manipulate their wine into the "international" style. But climate change is real and it is happening, and nowhere is it manifesting itself more clearly than in the world's vineyards. Again, the term *global warming* confuses people as they look out their windows and see rain and hail. It is better to use the more inclusive term *climate change*, which encompasses the extreme variations in climate that we are experiencing. The point is that now we have regions reporting their coldest, frostiest, snowiest, wettest spring on record, followed by their hottest, driest and summer and fall on record. Their yields receive a brutal reduction twice during the growing period. The only way that we will be able to adapt to these changes, and to ensure we have the time to adapt to them, is if they are universally communicated.

The First International Symposium on "Alcohol Levels Reduction in Wine" was held in Bordeaux in September 2013 and set out to assist the world's winemakers to do just that. In his "Welcome Letter," Christian Seely (general at Château Pichon-Longueville, AXA Group) wrote:

The increase of the alcohol level in wine related to climate change is one of [our major challenges]. This phenomenon observed all over the planet shows that grapes ripen more and more early, and would mainly result from global warming . . . It is now common to see quality wines with an alcohol by volume (ABV) of 13, 14 or even 15%. Since the eighties, each ten years, alcohol levels gained almost 1% with an average increase of 2 to 3%, if not more. This historical surge of ABV was measured in many countries: in the South-West of France, 15 years ago, the average alcohol level amounted to 11%; it now ranges between 13 and 14%. In Australia, the average was 12.4% in 1984; in 2004 it reached a striking 14%. In California, the average ABV was 12.5% in 1978 and soared to 14.8% in 2001.

The symposium looked at both adaptation and mitigation techniques. To tackle high alcohol levels you need both, but there is the concern that any adaptation of the wine after it has been made,—say, by adding acid, a well-known practice in warmer regions—can alter the wine's quality. It is never a good idea to fiddle too much with the wine's natural processes. It is agreed that trying to mitigate the alcohol levels in the vineyard is a more desirable approach. And this is what they were really looking at. The dialogue concerning changing regulations, introducing irrigation, changing varietals and even creating new ones, and how adaptation will change a wine's taste and the consumers' perception of this, is well under way.

ADAPTATION TECHNIQUES

CLONAL SELECTION: HELP VINES ACCUMULATE LESS SUGAR IN THE FIRST PLACE

In a paper submitted to the Oenoviti International Symposium on Alcohol Reduction in Wine, authors Serge Delrot et al. suggest that the best strategy is to "combine adequate vineyard management with the varietal and clonal genetic diversity of grapevine in order to decrease berry sugar content while maintaining the other properties required for making premium wines." They note that

> during the past decades, viticultural practices and climate change have resulted in an increase of sugar berry content at harvest. This results in wines which may have a too high alcohol potential. Although dealcoholization techniques exist and selected yeast strains may produce a lower alcohol content in the wine, it is less time-consuming and expensive to try to reduce the sugar content directly at the berry level. This requires a good knowledge of the

physiological and biochemical processes participating in the control of berry sugar content, and the design of modelling strategies and molecular markers.

(Serge Delrot, serge.delrot@u-bordeaux2.fr; Louis Bordenave1, bordenave@ bordeaux.inra.fr; ZhanWu Dai, zhanwu.dai@bordeaux.inra.fr; Alexandra Lusson, a.lusson@gironde.chambagri.fr; and Nathalie Ollat).

In other words, get it right, right at the start. They decided that the simplest way to avoid excess alcohol production during vinification is "to start from berries containing the right amount of sugars. This can be reached by combining adequate vineyard management with grape genotypes that naturally accumulate less sugar." I think that what they are saying is that in the past, in Bordeaux, for example, clones were selected to better deal with very different climate conditions at the time. Once upon a time, they wanted, they needed, clones with potential high alcohol and yields. In the past, most grape adaptations have been geared to trying to grow grapes in cooler climates, to maximize maturation. But now we find that our objective is the opposite. Now that the climatic environment has changed, so must the performance profile of the clones. This is the basis of all clonal experimentation: continually adapting clones to their environments and the changing needs of viticulturists.

In Bordeaux, the two grape varieties suffering the most from the warmer temperatures are Merlot and Sauvignon Blanc. They are already the two earliest-ripening varieties and so are now even more exposed to the increasingly hot summers and being harvested earlier and earlier. This is where clonal selection and experimentation is one of the primary adaptation techniques, whether with rootstocks or the scion. And it may seem incredible, but even in Canada new varieties are being selected to combat the heat: The sugar content of the Sabrevois variety rarely exceeds 21° Brix. "This variety exhibits medium vigor, low internode development, an average yield of 8t/ha and medium acidity. The Radisson variety (formerly called ES 5-17) does not exceed 20° Brix and has a low acidity" (Delrot et al.). Back to Bordeaux, and the Merlot again . . .

A good example is the work being done on Bordeaux's Merlot,

which is issued from the cross between Magdeleine noire and Cabernet Franc. It is one of the most famous and widespread varieties in the world. It occupies about 50% of the Bordeaux vineyard. Sauvignon blanc, which was formerly used only for the sweet white wines is the most planted white variety in the Bordeaux vineyards.

Twelve clones of Merlot are presently agreed, among which 10 are issued from a repository planted by INRA in Gironde in 1964. At this time, they were selected on the basis of their medium productivity and high alcohol potential. Improvement of viticultural practices now allows obtaining ripe berries more regularly, and low yields are desired to reach better quality. The clonal diversity of Merlot is presently revisited to select 20–30 genotypes out of a population of 256 genotypes present in the Bordeaux INRA repository. The screening is first based on the amount of berries and the shape of the cluster, and on early phenology. In the Bordeaux area, a lag phase of about 110 days and 45 days are observed between mid-flowering and maturity, and between mid-*véraison* and maturity respectively. Other measurements concern sugar content, total acidity, malic acid, pH, polyphenols, berry weight and berry tasting.

And, twenty clones of Sauvignon blanc are presently agreed, and there are 3 French repositories gathering 400 accessions, including one at INRA Bordeaux (Château Couhins). Five clones (108, 242, 316, 317, 905 et 906) have been particularly studied by the Chambre d'Agriculture of Gironde and exhibit an interesting variability in terms of yield, sugar content, acidity, phenology, resistance to Botrytis and wine aromas (Ibid.).

Clonal selection experimentation is happening in all wine-producing regions. In Tuscany, concerns over the Sangiovese grape resulted in the *Progetto Chianti Classico 2000* in which 12 leading wineries (with 14 vineyards over 25 hectares) located in Chianti Classico participated: Badia a Coltibuono, Fontodi, Castello di Albola, Lilliano, Rodano, Castello di Cacchiano, Selvole, San Felice, Isole e Olena, La Madonnina Triacca, San Nicola a Pisignano, and Castello di Rencine.

Sangiovese has had quite a checkered viticultural history. War and poverty meant that wine was a foodstuff needed for sustenance, and high-yield, low-quality clones were used at the expense of quality clones. Chianti was always a mess. Between the Baron Ricasoli (1850s) deciding that Chianti should be a blend of Sangiovese and Canaiolo (fine), but also 30 percent of two white grapes, Trebbiano and Malvasia, and the *mezzadri* leaving (the sharecropping system broke down) for the cities, Chianti was a very bad wine enshrined in Italian law, until the nouveau riche Italians zoomed into Tuscany via the Autostrade 1 from Milano and Roma and used their new money to make new wines . . . and break a few outdated laws. The Super Tuscans were illegal.

This historical hangover is still being addressed today. The clonal selection program was born through a desire to return Chianti to Sangiovese and to rediscover the true attributes of this grape. The Super Tuscans were made of the French varieties, Merlot and Cabernet Sauvignon, now so ubiquitous that they are "international." The Super Tuscans are not Italian. Chianti Classico, with its recent law allowing Sangiovese to be between 80 and 100 percent, is Italian once again. Except that we still have Chianti producers using Merlot and Cabernet Sauvignon and we still have Chianti Classico producers intent on making Sangiovese in the international style, which is being aided by the warmer climate.

Anyway, these experimental plantings of selected Sangiovese clones were monitored over a decade with micro-vinifications carried out from every single plot of clones from 1993 to 2000. In 2000, 180 presumed clones of Sangiovese were isolated, along with 54 clones of Canaiolo and 5 of Colorino. Out of this range of "presumed clones," twenty-two clones of Sangiovese, ten clones of Canaiolo, and two clones of Colorino were selected. Finally, further tests led to the homologation of four new clones of Sangiovese in the Catalogo Nazionale delle Varietà di Viti: CC 2000/1; CC 2000/2; CC 2000/3; CC 2000/4. The University of Milan recently released three new quality-oriented clones of Sangiovese as well, selected separately from Progetto Chianti Classico 2000. They are PF30, Janus10, and Janus50.

These clones have been used by these wineries ever since, and made available to other wineries, and the wine we are drinking today are most likely a result of this experiment to produce lower-yielding, heat-resistant Sangiovese. It never ends; in 2010, an international symposium was held in Maremma, at the Petra winery, where again the future of Sangiovese was discussed along with concerns over finding new planting areas—the Maremma coast, for one—and the impact of rising alcohol levels.

Clone	Source	Features
CC 2000 / 1	*Az. Agr. Rodano*	Clone reputed as very promising because of the loose shape of the bunch, moderate acidity, early ripening profile, and high content of anthocyanins. Suitable for producing "premium" Sangiovese wines meant for medium- to long-term bottle-aging. Good resistance to all main vine diseases.
CC 2000 / 2	*Isole e Olena*	Clone reputed as promising because of the loose shape of the bunch, early *véraison*, good varietal aromas, and high alcohol content. Micro-vinifications scored high levels of total polyphenols, medium to high acidity, and notably high alcohol content. With thick skin, medium-low vigor, phenolic ripeness occurs the first ten days of October. Average resistance to all main vine diseases. Suitable for producing "premium" Sangiovese wines meant for medium- to long-term bottle-aging.

(*continued*)

Clone	Source	Features
CC 2000 / 3	*Isole e Olena*	Micro-vinifications scored average levels of total polyphenols, medium/high acidity, and high sugar accumulation/high alcohol content. Phenolic ripeness occurring mid- to late September. Good vigor and good resistance to all main vine diseases. Suitable for producing medium-bodied, tasty Sangiovese wines meant for medium- to short-term consumption.
CC 2000 / 4	*Castello di Ama*	Clone reputed as promising because of the small-sized, conic-shaped and loose cluster, because of consistent vegetative vigor and good content of total polyphenols and high acidity. Phenolic ripeness occurs mid- to late September. Intermediate vigor and average resistance to all main vine diseases. Micro-vinifications scored notable levels of total polyphenols, high acidity. Suitable for producing medium-bodied, perfumed, and varietal Sangiovese wines meant for medium- to short-term consumption.

Toscana IGT Sangiovese, Luca Mazzoleni.

In Verona, using the Corvina grape (one of the grapes used in the Valpolicella blend), a team of Italian geneticists identified genes that help protect the fruit from the vagaries of the weather and could serve as a platform "for breeding new cultivars with improved adaptation to the environment" (Genome Biology). The team grew the grapes in eleven vineyards across the Verona region and harvested berries at various stages of ripening for three years, to analyze which genes were expressed under what conditions, finding genes, for example, associated with a wine's taste, color, and mouth-feel (John Roach, NBC News).

In Spain, the Torres estate is experimenting with new rootstocks that show increased resistance to drought, or that delay ripening to allow the aroma more time to mature. They, too, know that often, it is the past that becomes our future. They have rediscovered and are using twenty forgotten Catalonian grape varietals that they believe will fare better in high temperatures than common varietals imported from northern Europe. This harks back to our discussion on indigenous versus traditional, doesn't it? It is no mystery that we are finding ancient road maps and following them. We'd lost our way and headed down the international branding, "make it easy for the Americans to pronounce," easy-money highway.

As we said, there are no easy ways out. When you mess with nature, there are ramifications. So what is the flip side of clonal adaptation? John Roach quotes Gregory Jones explaining how new climate-resistant strains of each grape varietal will affect the wine itself. "In an ideal world, of course, you would never breed out beneficial or quality characteristics while you're trying to breed 'in' environmental defenses," Jones says. "So there is clearly a balance that has to be done there." *Balance* being the key word. In Australia, soil salinity is a huge problem, and so experts are breeding vines with high salinity tolerance. Isn't clonal breeding just a Band-Aid, a quick fix for the larger problem? How much tweaking will ever be enough?

REDUCED TILLAGE

In her study on "Vineyard Cover Crops and Tillage Practices," Dr. Kerri Steenwerth has determined that tillage is one of the largest viticultural activities that produces greenhouse gases. And happily, it is the practice that is the best understood, and the most easily controlled and measured by grape growers. Tillage, or plowing, is meant to loosen and aerate the top layer of soil, which facilitates planting the crop. It also helps mix harvest residue, putting the organic matter and nutrients back into the soil, as well as destroying weeds. Tillage is a great weed control method. By disturbing the top few inches of soil around the crop plants but with minimal disturbance of the crop plants themselves, tillage kills weeds by both uprooting them and burying their leaves, which cuts off photosynthesis. Apparently it is even better to till at night, when the weeds are dormant.

As with all practices, balance is key. Too much tillage, or aggressive tilling, dries the soil before seeding in hot climates (in wetter climates it helps dry the soil out), erodes the soil's nutrients, reduces its ability to store water, and results in soil compaction, which increases water and chemical runoff. Reducing tillage reduces soil erosion, runoff, dust pollution, and greenhouse gas production. Reduced tillage encourages healthier soil biology and increases organic matter.

Conventional tillage leaves less than 30 percent of crop residues on the surface and requires multiple passes with the plow. It robs the soil of carbon, creates more CO_2, and uses the most fossil fuels. Conservation tillage leaves more than 30 percent of crop residues on the surface, allowing more carbon to enter the soil's organic matter; less CO_2 is released, and less fossil fuel is used.

No-tilling means that there is no disturbance of the soil surface. It allows the most carbon to enter the soil's organic matter, and reduces costs and environmental change by reducing soil erosion and diesel fuel usage. Crops can be grown for several years without any tillage through the use of herbicides to control weeds, but then the synthetic herbicides and fertilizers increase N_2O production.

CANOPY MANAGEMENT

Canopy management is the blanket term for the practices of pruning, trimming, shoot thinning, leaf removal, and trellising vines—the methods by which the amount of foliage is controlled, varying the amounts of air and sun that reach or don't reach the grapes. Better trellising techniques should lessen the amount of the pruning and thinning needed. This issue, too, is rife with debates from both sides of the world. And again, there have been books written on this vast subject alone. The trick with canopy management is to achieve a balance between foliage and grape distribution. Enough foliage is needed to stimulate photosynthesis, thus providing the sugars, but not so much that the sunlight is blocked from the grapes and they do not ripen. And an overshaded canopy can lead to smaller yields with grapes that have an herbaceous, stemmy, stalky taste. A more open canopy has the advantage of allowing more air ventilation, which helps protect against diseases such as rot and mildew, and trellising the vines keeps them off the ground, safer from frost damage, where applicable. Another variable in the equation is planting density, or row spacing. The Old World has traditionally had tighter spacing, so they were managing the canopy anyway—simply for spacing issues, favoring open canopies.

The Old World stance is that the vine has to struggle and yields have to remain low to ensure quality: Vine vigor is an enemy. The New World approach is that vigorous vines can give high yields of high-quality grapes so long as they are managed. Canopy management was a primarily New World concern and it came to the forefront as a way to deal with excess vigor—which was not an issue in the low-vigor vineyards of the classic wine regions of Europe. With the European climate shifting to mirror that of the New World, vigor is increasing, and adapting canopy management and trellising techniques is now something the Old World is looking at very differently.

IRRIGATION: A QUESTION OF TASTE, AND WASTE

This is a yet another huge topic. Ultimately, all practices in the vineyard will eventually play a part in the taste of the final wine, but none so much as the practice of irrigation. There is a huge debate over this between the Old and New Worlds, naturally. The Old World camp has always said that irrigation dilutes the wine and that better fruit relies upon rainfall—but they had the climate that allowed them to say that. The New World producers have always said, "If a tomato needs water, you water it. Grapes are the same"—but then, they have a climate that forces them say that. In a nutshell, *irrigation* is a wide term, encompassing a variety of practices according to the amount of water used and the frequency with which the water is applied: from flood or furrow irrigation, to spray irrigation, and to drip or trickle irrigation . . . and from the first day of the growing season through to harvest, or once a week, or once a day, or continuously . . . Unirrigated vines are forced to dig down deep to find moisture, and they pick up nutrients through the soil formations as they do so. Irrigated vines often miss out on vital nutrients because their root systems remain on the surface, where the moisture is. So the produces make it even easier for them—they add fertilizers to the water in the drip irrigation system (called "fertigation"!). So the vines are fed and watered without even having to get out of bed—literally. All that they need is home-delivered directly to them. They are lazy. And like lazy, spoiled children, they will not grow up into very interesting adults, will they?

With heat erasing varietal character and soil influences, and irrigation diluting it, good wine—forget *fine* wine—doesn't stand a chance. Irrigation also is the most damaging and wasteful viticultural practice. Although Carmel Kileline MW points out in her dissertation that "while 99% of the water used in wine-making is used for irrigation rather than in the winery, grapes are still a relatively modest user of water. In Riverland (Australia), 290 litres of water are used per 750ml bottle of wine. Rice, by contrast, requires 2,380 litres per kg, and cotton, 5,020 per lg pf cloth." So perhaps wine producers need not feel too enormously guilty. But irrigation remains our biggest dilemma, both in terms of affecting wine's quality and taste, and in terms of conservation practices. Does a dry region keep increasing its irrigation until it runs out of water? At what point should a region change crops or consider other agricultural uses?

Growing grapes is growing fruit. The basic principles of gardening and fruit farming provide the needed guidance. Any good gardener will tell you that overwatering through either rain or irrigation dilutes fruit flavor and increases yield. This is the first thing one is taught in "wine school." It is part of the Wine 101 curriculum, and it is embedded in the European psyche. It is illegal to irrigate in Europe, and for good reason: You *can* taste the difference. They didn't just make up the rule because they felt like it. If they thought irrigation was best for the vines, or that there was an easier, less expensive way to water their vines, they would have done it. What's interesting is that now that they are experiencing heat and drought in New World proportions, they suddenly are saying, "Well, a bit of drip irrigation here and there won't hurt."

But it can. Remember, irrigation allows the vine to be lazy; the roots stay in the top 40 centimeters of soil and don't seek out the water or nutrients in the subsoils and sub-solum. As Dr. Emmanuel Bourguignon states: "Permanent irrigation leads to a shallow root system. You get a really big mat of fine roots in the first 40 to 50 centimeters of soil. The most fertile horizon in the soil is in that first 40 centimeters because that is where you have the organic matter. If your roots stay in that horizon you will end up with some slight vigor problems."

This increased vigor, or vegetative growth, creates a large canopy, which is particularly problematic in sunny climates because "you end up getting massive photosynthesis—you just end up with a high level of sugar and your alcohol potential is high," says Bourguignon. "So you dilute the *terroir*, but you tend to increase the varietal character. You can have a good canopy and make a good varietal wine." Think of a New Zealand Sauvignon Blanc and its loud, cartoon-like varietal profile: It *screams* "I am Sauvignon Blanc." This is exactly the result to which Dr. Bourguignon is referring. There is nothing wrong with that if it's the result you want—if you are "making an entry-level fruity wine, but you can forget about minerality and sense of place." If you want to be unique, however, irrigation will make that very difficult. (Rebecca Gibb, "One of the World's Leading Soil Experts Tells Wine Producers to Turn Off Their Irrigation," February 5, 2013. Dr. Bourguignon and his family have worked with clients all over the world, including Domaine de la Romanée-Conti and Domaine Dujac in Burgundy, Domaine Huet in the Loire, Vega Sicilia in Spain, and Harlan Estate in the United States.)

Another huge problem presented by irrigation is the increase of the soil's salinity, which harms the grapes. In Europe, where the soils (*terroir*) are King, and are so much a part of a wine's composition, changing that very composition will change the taste of its fruit. Salt buildup is such a problem in some southern Australian vineyards—vineyards that have had to rely upon heavy irrigation since their inception—that the winemakers have had to abandon them. Another example of how manipulating the environment eventually catches up with you. Ideally, irrigation would perfectly mimic the effect of rainfall, with a heavy "deluge" in winter or early spring, as long as the soil is not so parched and compact that it cannot store or hold the water adequately. This is often the case when the rest of the season is not humid enough and any moisture in the soil is evaporated. Such copy-cat actions would simply mean that the water sits on the top of the soil and causes problems.

In his article "The Dangers of Soil Salinity," Tim Teichgraeber examines the problem of salinity in the United States. He quotes biochemistry professor Grant Cramer of the University of Nevada–Reno: "Anywhere you have arid climates, you're going to accumulate salt in the soil. Australia has significant salinity problems, and I would imagine some of the North African grape-growing areas have significant salinity issues too. It's a worldwide problem. Certainly the San Joaquin Valley would also have problems." Teichgraeber explains that when salt levels get high enough in the vine, the leaves start to display "leaf burn" or browning, as they do with some other vine afflictions like Pierce's disease. Another salt accumulation problem is caused by the way salts change the structure of the soil itself, and the effect that this has on plants. Salt also changes the way the roots grow. "Salts are more than just the sodium chloride you might use to garnish your margarita or make your strip steak really pop. They're a whole class of ionic compounds made up of positively charged cations and negatively charged anions that are neutral when combined. Among the potentially phytotoxic salt components are sodium, chloride and boron, all of which can cause crippling decreases in vine vigor or even vine death at elevated levels."

Irrigating saline soils with salty groundwater tends to exacerbate the problem and cause dehydration—the most basic salt accumulation problem for plants. "Another tactic is drenching irrigation sufficient to wash out the accumulated salts and restore a healthy balance to the soil. Fresh water rainfalls are integral to keeping saline soils in check in coastal regions. Inland regions like Nevada or Texas often have more serious salinity problems because they seldom see a good drenching. In Nevada, we don't have rivers to wash the salt out into the sea," Cramer continues. "If it's possible to flush the excess salts out of saline soils with mountain runoff, there are parts of Western Nevada that might flourish as grapegrowing regions just the way Mendoza has in Argentina, provided that the mountain runoff is used efficiently. Both are desert regions on the leeward side of mountains that see substantial rainfall. It may be possible to wash the salts out of previously barren soils in areas with adequate drainage."

So perversely, the best way to cope with soils that have high salinity is to flush them out with huge doses of fresh rainwater. But if these places had enough fresh rainwater with which to flush out and drench their soils, they wouldn't need to be irrigating in the first place. I know that I am overgeneralizing a bit here, but when you look at this with some perspective . . .

REGULATED DEFICIT IRRIGATION (RDI)

There is much experimentation centered on irrigation techniques, and most of it leads to using less water, less frequently. Even in areas where irrigation is employed, it is acknowledged that too much compromises grape quality and makes fine wine impossible.

In the inland areas of the US Pacific Northwest, for example, rainfall averages only 4 to 12 inches per year, and "the arid conditions have not been conducive for vineyard owners who produce and market high-quality wine grapes." The adopted irrigation practice has been regulated deficit irrigation; more than 60 percent of the wine grapes grown in Washington are grown using this method, which was "borrowed from the peach industry and modified for use with wine grapes." By withholding water during the period between when the grapevine first sets fruit and *véraison*, the vegetative growth is controlled and the size of the grape berry is reduced (which means more concentrated wines).

But concerns have arisen over problems that can cause over-irrigation and compromise the grape quality. An article from *Science Daily* states that researchers at Washington State University have pinpointed a problem with the ability to accurately measure soil moisture with this system, The "zone" being measured was the area "right below the drip emitter," and their trials determined that a more accurate measurement is taken in the zone that is within a twenty- to forty-centimeter radius of the drip line emitter. This seemingly small detail should go a long way toward avoiding over-irrigation.

In Larry Williams's paper on irrigation and California wine grapes (Department of Viticulture and Enology, UC Davis), he researches how to determine how much irrigation water is required to grow quality wine grapes, stating that it depends on the site, the stage of vine growth, row spacing, the size of vine's canopy, and the amount of rainfall occurring during the growing season. He observes that

> coastal winegrape production areas in California are character-
> ized by warm days and cool nights, although high temperatures
> (104–116°F) may occur for a few days each growing season. Some
> areas may have fog lasting late into the morning. Rainfall is greater
> in northern coastal valleys and diminishes as one travels south. In
> coastal valleys, evaporative demand can range from 35 to 50 inches
> of water throughout the growing season (between budbreak and the
> end of October). Many of the soils in the coastal production areas
> are clay loam to clay-type soils, which hold more water than sandy-
> type soils. Since the majority of rainfall occurs during the dormant
> portion of the growing season in these areas and vineyard water use
> can be greater than the soil's water reservoir after the winter rainfall,
> supplemental irrigation of vineyards may be required at some point
> during summer months.

When producers are deciding when to begin irrigating, they can do this in several ways: With soil-based tools such as a neutron probe and ca-pacitance sensors, which can determine the actual or relative amounts of water in the rooting zone of grapevines. Or with plant-based tools, such as a pressure chamber, which can be used to measure vine water status. The information needed to schedule irrigations at daily, weekly, or other intervals throughout the growing season includes potential evapotranspiration (ETo) and reliable crop coefficients (kc). Potential ET (also known as reference ET) is the water used per unit time by a short green crop completely shading the ground. Ideally, the crop is of uniform height and never water-stressed. ETo is a measure of the evaporative demand of a particular geographic region throughout the year. Current (or near-real-time) ETo data are available from the California Irrigation Management Information System (CIMIS), which is operated by the California Department of Water Resources.

There are more than ninety weather stations located around the state where environmental data are collected to calculate ETo. Environmental variables measured to calculate ETo are mean, hourly solar radiation, air temperature, vapor pressure, and wind speed. These variables are then used to calculate other variables such as net radiation and vapor pressure deficit, which are then inserted into an equation to calculate ETo. Potential ET can also be obtained from weather stations operated by other entities (such as stations operated by the Paso Robles Vintners and Growers Association in the Paso Robles region).

Williams concludes that "initiating vineyard irrigation later in the growing season or the use of deficit irrigation may restrict excess vegetative growth, whether for grapevines grown in the interior valley of California or along the coast. This would minimize the cost of canopy management for vines that, in the past, became too vigorous due to excessive irrigation." Grapevines can be deficit-irrigated at various fractions of estimated full ETo with minimal impacts on yield but with a potential to increase fruit quality. Thus, in most cases, one may not have to apply water amounts that meet or exceed the estimated vineyard water use requirements presently acknowledged.

Scientist Richard Gawel provides an Australian perspective when he takes us back to a less technological time, remembering "the old timers who sought out sites that ensured trouble-free grape growing and decent yields. Over time, their old vines negated the need for irrigation by planting their roots firmly down a meter or three which has enabled them to access some moisture even in the driest of times." He muses over the fact that as a wine taster, "there is no doubt that these old unirrigated vines can make stunning wines with a denseness and richness that are unsurpassed." But today, nearly all Australia's vineyards rely upon some form of supplementary irrigation—the climate demands it because even in the coolest and wettest regions, evaporation "outstrips" the natural rainfall during the growing season. He explains that even in the Lower Hunter, where the summer rainfall is very high, even with a higher average than Bordeaux, because the annual rainfall is varying so extremely, the vines are "considerably water-stressed."

Gawel acknowledges that over-irrigating produces a wine that is low in alcohol and color, and has a thin, weedy, and dilute palate. "You come across these regularly in wine tastings. They certainly stand out at least in their mediocrity." He continues, "However, it is equally true that vines that are highly water stressed can also produce poor quality grapes with much lower colour, flavour and tannin than vines that have experienced low or moderate levels of water stress."

This is the truth we all know: Moderation is everything. Moderately stressed vines will produce the best wine. Because they "produce less vegetative growth than their over-irrigated counterparts and this results in vines that are balanced in terms of the number of leaves required to ripen the fruit, and ensures that the vines canopy is not overly shady. Dense, shaded canopies are the cause of unattractive pale, thin, and acidic herbally flavoured wines." But in Australia, even achieving moderate stress requires irrigation.

Apart from reduced deficit irrigation, as described above, Gawal reports on another irrigation approach called Partial Rootzone Drying or PRD. He describes it as "tricking" the vines into thinking that they are stressed by watering only one side of the rootball at a time. By doing so, the vines slow down their lateral shoot growth (decrease their vegetation); yields are not affected and the amount of irrigation water needed is nearly halved. A serious advantage, he writes, especially in places like the Murray Darling Basin. But others report that this technology is still being researched: PRD may not be as effective as is hoped in regions with excessive evaporation or a heavy demand for supplementary irrigation—that is, in the places where we most need an effective irrigation system.

WHAT IS DRY FARMING

Alastair Bland's article "To Grow Sweeter Produce, California Farmers Turn Off the Water" reports on commercial growers in California who are turning to "dry farming," and how, throughout the state, this "unconventional" technique seems to be catching on among small producers of tomatoes, apples, grapes, melons, and potatoes. This is hysterical. It refers to the practice of relying only on natural annual rainfall for growing grapes—which is what you are supposed to do in the first place. Dry farming is what Europe does. But the clever growers in the drier regions, crippled by water scarcity, have figured out a way to turn it around and make it a trend. It is a bit like when, in the 1990s, the Californian winemakers figured out that Merlot, Cabernet Sauvignon, and Cabernet Franc taste better when they are put together in the same bottle and then called it "Meritage," as if they had come up with the idea themselves. Paul Vossen, a University of California farm adviser in Sonoma County, also smells a rat and says of the dry farmers, "They do it because they have to, and so they'll make it part of their marketing strategy."

Dry farming does preserve water resources, good. But it also means that you are going to get smaller, denser, sweeter fruit. Good? For all fruit other than wine grapes, probably so. Too little water leads to raisins, and too much leads to dilution—try eating a hydroponically grown tomato.

At Whole Foods Market in Sebastopol, about 50 miles north of San Francisco, dry-farmed tomatoes have become a shopper attraction, according to produce buyer Allan Timpe. "People definitely come here to get them," he says. "Once someone tastes a dry-farmed tomato, you've got a customer for life." He also carries locally grown dry-farmed potatoes, which he says "are dense and really flavorful."

Dry farming in sandy soil, through which water drains easily, doesn't work. Just as important, the plants, or trees, must be dry-farmed from the time they're planted. "We get the plants going with a little water, then cut it off after a few weeks," says Kevin McEnnis, who dry-farms Early Girl tomatoes at Quetzal Farm in Santa Rosa. Then, he explains, as the young but quickly growing vines become thirsty, they send their roots deeper underground than they otherwise would to find moisture, which can remain in the soil all year.

A technique that also helps dry farmers lock water underground is frequently tilling the top foot of soil into a fluffy dust layer. Underground moisture that creeps upward through the earth cannot break through this layer, and it remains below the surface. But, McEnnis adds, "Farming without irrigation has a major drawback, it dramatically reduces yields." Stan Devoto, a farmer in Northern California, says his dry-farmed trees produce twelve to fifteen tons of apples per acre per year. Irrigated trees, on the other hand, may bear forty or fifty tons. And McEnnis says he harvests about four tons of tomatoes off his acre of vines each summer and fall, whereas conventional growers may reap forty (Bland).

Do the world's fruit and vegetable farmers and the wine growers ever talk to one another? Winemaker Will Bucklin, of Old Hill Ranch in the Sonoma Valley, has an underground water supply. Still, he dry-farms fifteen acres of old-vine Syrah, Zinfandel, and other varieties; he's one of just a few California winemakers currently dry-farming. His grapes are smaller, and yields at harvesttime slightly lower, than on irrigated vineyards. Alastair Bland continues to quote Bucklin as he says: "But the small size of the berries means there is a lower juice-to-skin ratio." And he adds that since grape skins contain flavor-making tannins and polyphenols, dry-farming can, at least in theory, produce richer, more intense wines. Bucklin concedes that "he isn't certain that a novice taster, or even an expert one, could tell a dry-farmed wine from a conventional one in a blind tasting." Now how to explain that a dry-farmed wine *is* a conventional one, at least per the European model? Is he suggesting that irrigated vineyards are the new "convention"? Interestingly, most of Napa, including all the classic founding wineries, was dry-farmed until the 1960s, when overhead irrigation was introduced to the valley. At that point, irrigation was not about relieving heat stress or drought conditions; but rather, it was being used as a method of frost prevention. This is further evidence of how much the climate has changed. Of course, it also didn't hurt that irrigation increases tonnage and hence, profits. Otherwise, why would they have stopped the dry-farming method if it were indeed producing superior fruit? Because they couldn't. Soon after becoming accustomed to the advantages of irrigation, they found that it eventually became a necessity. But dry-farming works best in cooler climates. So while dry-farming is a boon for other Californian fruit crops, will it work for wine grapes already being grown in a hot climate? You can taste the difference between a wine made from diluted, irrigated grapes and one made from overly ripe and heat-stressed fruit. It seems to me that they are now only choosing the latter because they have no choice. But once irrigation takes hold, can a winery go back? Once you have subjected your soil and root systems to irrigation, assuming that it was the right sort of soil to begin with, it is hard to bring it back, regardless of how much tillage and mulching is employed. It will be interesting to see if and how this "trend" progresses.

IT'S IRRIGATION OR NOTHING

Playing the devil's advocate, proponents of irrigation hold that during the past decades, New World wine production has reached higher levels of quality and production consistency (not my words); the effects of global warming on *terroir* have become increasingly severe; and European viticulturists are reexamining traditional growing methods.

The latter is true. Very quietly, so no one can point at producers and laugh, irrigation has now officially been legalized in France. The droughts in 2003 brought southern French vineyards to their knees, and the result was a slight relaxing of the INAO ruling on irrigation. A decree dated December 6, 2006, permits irrigation from June 15 up to the French national holiday of August 15, although for AC vineyards it is allowed only if sanctioned by the individual appellation authorities. This had been brewing for a while. After the disastrous 2003 vintage, the Languedoc and Bordeaux vignerons forced Sarkozy to allow them to irrigate and to pledge to an ongoing climate change adaptation program. It was a case of irrigating or losing an entire year of wine. The new rules have been conceived in that very French way and do very little to give the needed control to the vignerons, who are the ones who actually know what is going on. For some reason, the drip-irrigation system is specifically not allowed, when it is far more efficient than spray irrigation. And the paperwork, permits, and permissions for each parcel, each grape, each region, and each quality designation are just mind-boggling. This really changes the game plan. This repositions some major Old World traditional regions in the same position as the New World.

So, in order to secure "long-term investment in *terroir* branding," to ensure stable vintage quality and ultimately compete in the mid- to high-level global wine market, and to combat climate change, more and more Old World growers are turning to modern irrigation methods. Another market trend has been created as European bulk wine producers were looking to move into the lucrative mid-level quality market currently dominated by New World producers. Bulk wine producers know that "modern irrigation methods can assist them in consistently producing the quality yield needed to effectively build label brands" (Netafim brochure). So basically the Europeans, who were making the best wines (and some of the worst, thank you), looked at the global market pyramid, saw that they were indeed sitting at the top and the bottom segments, and asked themselves if they could be part of that huge segment in the middle—the mediocre wines. "Hey, the world wants mediocre? We can do that, after all, and our mediocre is their fine wine."

Another anti-irrigation argument is being fought by many Portuguese winemakers in the Douro, who contend that there was never a need for irrigation there for three hundred years and that those who are using irrigation to increase their yields are creating a vicious cycle: Again, by irrigating, the root systems remain shallow (they don't need to dig deep for nutrients or water—they are being hand-fed), and these shallow root systems are thus more susceptible to drought. This may have been the case in the past, but the other traditional European regions, such as Languedoc-Roussillon, have not been irrigating, and they have been using viticultural practices that result in deep root systems—yet it still got too hot for them and they needed to irrigate.

Irrigation means considering the vineyard soil type and depth, climatic conditions and resultant evapotranspiration levels, the soil's water-holding capacity, the winemaker's preferred planting density and spacing, the vineyard's contours, and the agro-technical use, such as harvesting machinery and the like. And the consensus is that drip irrigation is the most efficient system; within drip irrigation, the "on-the-wire" method is the most common, used when weed control is accomplished by plowing since no laterals rest on the ground. The "surface" method is the simplest, most cost-effective, and most versatile form and is used when weed control is chemical under the vine and drip lines are laid on the soil surface under the vine along the trunk line. The subsurface method is the most aesthetic, reliable, and efficient technique because the laterals are buried twenty-five to fifty centimeters under the soil. This means that they are protected from weather and mechanical machines, that evaporation water loss is reduced by 10 to 20 percent as compared with a surface drip, and that the water is applied precisely to the center of the root zone.

Unlike conventional irrigation methods, drip irrigation does not consist of utilizing the soil as a water reservoir, since the horizontal water distribution is limited and depends on the hydraulic conductivity properties of the soil. Therefore, this method requires frequent irrigations in order to eliminate water percolation to soil layers below the major root zone. Irrigation frequency is determined by three major factors: soil properties, evapotranspiration rate, and water emission rate of the irrigation system.

Drip systems are normally designed to supply the water requirement of at least one day during the top season. At the beginning and at the end of the season, when the water requirements are low (due to low evaporative demands and low canopy surface irrigation), frequencies may vary between one and two per week. During the major season, frequencies may increase to daily or every other day. Sandy soils require more frequent irrigations than heavy soils. Accumulation of large deficits may require extended duration of irrigation hours and may result in water percolation or runoff.

OUTSIDE AIR-CONDITIONING?

There is a new, very low-impact irrigation method being looked at. An Italian company, Oasiclimatica, has developed a system called cooling dew which "air-conditions" the vines and reduces the berries' risk of thermal stress. A fine mist is sprayed between the rows at short intervals, cooling the grapes' environment. This technology lowers the temperature, using a minimum of water, without creating humidity. As soon as the system's heat sensors detect an air temperature over 86°F (30°C), it automatically sets off a high-pressure pulse of fine mist that decreases the temperature in less than one minute. It does not mist the leaves or the berries, which would risk creating an environment for disease. Two and a half liters of water at about 100°F (38°C) are needed to decrease the temperature below 86°F in less than thirty seconds. Lowering the temperature of the air is meant to stop the risk of the inner temperature of the berry increasing. In 2013, tests were performed in Umbria (Orvieto), the very hot center of Italy, on two neighboring parcels, one equipped with cooling dew and the other, the control parcel, without. They showed that even with an exterior temperate of 111°F (44°C), the temperature in the vines did not surpass 83°F (28.5°C).

At the other end of the spectrum is the use of the ancient system of flood irrigation: extreme irrigation. The Argentinean wine association writes in "Growing Grapes in a Desert" that "any visitor to Mendoza, or indeed most of Argentina's north-west interior, will be astonished that it is possible to grow anything at all, far less world-class wine. The entire region is in the very substantial rain-shadow of the Andes mountain range. The result is a dusty, dry countryside where not only does it practically never rain (there is a total average of less than 200 mm of precipitation per year), but the air is generally throat-parchingly dry too, with relative humidity levels between 40% and 70%." Now, I would have to take issue with the term *world-class wines*. Yes, the wines produced there are capable of filling a place in the international wine supply, but they are not fine wines.

Here the ancient irrigation system was inherited from the indigenous Huarpes people, who developed and built a complex and sophisticated system of irrigation channels to bring water from the Mendoza River to the arid plains. These channels featured advanced hydrodynamics techniques that allowed the regulation and control of the flow of water, allowing efficient use of the scarce resource (the Mendoza River being fed almost solely by snow-melt in spring and summer). The system supplies both Mendoza's residents and all its viticulture with water. Irrigation channels have been extended and added to, but the system remains the same. Water is rationed between vine-yards and farmers through the opening and closing of miniature flood control gates. Once in the vineyards, the growers use the same technique to flood irrigation channels around the base of their vines. This "simple flood-irrigation technique has been used for centuries in Mendoza and is only now, in a very few cases, beginning to give way to more complicated and expensive (if more efficient) drip irrigation systems."

The association goes on to state that its "desert climate and advanced irrigation system gives Mendoza's grape growers a unique advantage. With complete control over the watering of their vines, and in combination with the hot daytime temperatures and cool nights during the grape ripening seasons, conditions are almost ideal for growing grapes with ripe, intense fruit characteristics and good acidity levels." What this is really describing are the conditions that produce wines that are easy-drinking (diluted), boring, and commercial, with exaggerated varietal characteristics. What I call the "cartoon character" wines. Extreme irrigation: Is this the future?

HOW MUCH LONGER IS IRRIGATION VIABLE?

For those who see irrigation as the way forward, it, too, will only create a sort of holding pattern until it, due to lack of water supplies, is no longer an option for some regions. While irrigation makes wine production possible in some regions, it is creating a greater demand on our diminishing water sources. While many wine producers are looking at ways to reduce their water usage in the winery, they continue to irrigate. We've seen the New World wine-producing regions tout their sustainability credentials and sign up to the carbon-neutrality craze, but still, large swaths of the industry cannot survive without irrigation. Add to the problem the huge amounts of energy required to pump the water. How much longer is this going to be a viable way to grow grapes?

Returning to the possibility of dry-farming, Dr. Bourguignon believes that vines can manage quite nicely without water. "There are vineyards in Turkey and Lebanon that get less than 400mm of rain a year and are thriving without irrigation," he points out. "Vines will do as little as possible to reach a water source. Make it tough for them and they will work harder. The vine will naturally shift resources. If it doesn't get a lot of water it will use more energy on root growth to get to water."

But what to tell to the winemaker in Coal River Valley, Tasmania, who had three months of 90 millimeters of rainfall and 580 millimeters of evaporation, who would not have had a crop without irrigation and who insists that using supplementary irrigation with regulated deficit irrigation from berry set to *véraison* can achieve concentrated grapes without dilution. Is that sustainable? Or do we have to keep drinking their inferior wine so that they don't go out of business? Or until they run out of water? If an environment has to be contrived to such an extreme extent that every step of the growing process is altered, assisted, added, supplemented, reduced, researched, selected, or engineered . . . it may be time to pack up. Painful choices will have to be made, and a lot of people are going to be out of a livelihood.

THE ARGUMENT FOR IRRIGATION

Is there an argument for irrigation? It seems that the only argument is that when faced with the choice of irrigating and having a viable crop, and not irrigating and not having a viable crop, then irrigation will be chosen, even by its harshest critics. The New World–Old World divide is closing, and the move to advanced irrigation in traditional grape-growing regions reflects the need to adapt to the world's climatic changes. We know that modern irrigation systems promise reduced water expenses, reduced unwanted vegetation growth, and increased yield quality; they attempt to guarantee quasi-consistent wine vintages. Irrigation will soon be firmly a main ingredient in Europe's winemaking recipe—albeit a recipe they may not wish to pass down to future generations.

And irrigation can in no way be relied upon as a sustainable solution. Already every other agricultural crop relies upon it. Forget the quality considerations; the world is trying to grow food in deserts. Wine will simply have to take a backseat. Look at Dubti, Ethiopia. Miles of industrial irrigation, canals, and diversion dams have transformed 120,000 acres of desert to quench the multimillion-dollar Tendaho sugar plantation, an Ethiopian-Indian project that could make Ethiopia the sixth largest sugar producer in the world, thus breaking the country's need for foreign aid. But at what price? Despite the employment of fifty thousand migrant workers, others have been driven from their lands, or the diversions have meant that the Awash River levees are dry. When the water dries up, what then?

Irrigation will always be a huge topic of debate. For me, there is no "better" system than another. Once you have to begin irrigating, a dependency is being created that "starts to supplant natural root distribution"—and the ride down the slippery slope of compromise begins. This is a path the commercial, international wines know well. But they can afford to. Their investment/profit ratio can withstand the compromises. It is the fine wines, those that rely upon the unique personality profile of their soil bed, that will only find their demise at the end of the path. "High prices for these depend on their uniqueness and site typicity. These characteristics depend on an intimate relationship among roots, subsoil, underlying geology and climate that cannot be reconciled with any major need for irrigation" (John Gladstone, *Wine, Terroir and Climate Change*).

But let's really get our thinking straight on this . . .

WATER OR WINE?

After years of unrelenting drought, Phil Mercer writes in his article entitled, "Australia's Wine Region Threatened by Drought," (2008), that the lush valley soils of southeast Australia's Riverina wine belt were becoming cracked and dry, putting the competitiveness of best-selling labels such as Jacob's Creek at risk. But that seems like such a trivial worry when "more than 10,000 families have been forced off their land"—sheep and wheat farmers. Forget grape farmers. And wine, really.

"Worst hit have been the creeks and streams of the Murray-Darling river system, where around 1,300 growers produce more than 400,000 tons of mainly Shiraz, Cabernet Sauvignon and Chardonnay grapes, about a quarter of Australia's total"—all for those hugely commercial, bland, and ghastly labels such as Lindemans and Roberts Estate. These vineyards have "relied on highly regulated irrigation systems flowing from enormous reservoirs in the nearby Snowy Mountains," the article reports. But storage levels have fallen to such an extent that there are fears "that the diminishing water flow could lead to a reduction in grape yields." Astonishingly, the winemakers have "escaped the worst of the crisis" because "their business is so important to the local economy that it has been guaranteed water." Is this not a most sobering thought? Where are our priorities? They ensure the wine companies get the water and not the livestock or grain farmers . . . the *food* producers.

He continues:

> But for other inhabitants of the Murray-Darling basin, a vast expanse of land the size of France and Germany combined, the consequences are much more serious. Wagga Wagga, a rural community in the Riverina district, is facing disaster. "The situation's extremely tough," said Alan Brown, a livestock and wheat farmer. "Our season this year is hanging on a knife edge. If we get another crop failure I'm not sure what will happen, but it won't be pretty. The problem with drought is you're accumulating debt all the time," said Les Gordon, the president of the Rice Growers Association of Australia. "Farmers close to retirement are seeing their nest egg disappearing, their ability to retire starting to diminish and they are a long way behind where they were 10 years ago.

Apparently, the suicide rate among farmers is double that in the wider Australian population. Mercer reports that in the Riverina town of Griffith, 350 miles west of Sydney, the mayor, Dino Zappacosta, said health authorities were establishing a twenty-bed mental health unit to cope with the fallout. He quotes Kathy Maslin, who farms sheep and cereal crops on five thousand acres near Beckom in the Riverina: "In recent years we've had very low income or no income, so things have been stressful. For most people in our area, if we don't get a harvest this year, they'll be looking to find any way they can to get out." Well, someone pass her a chilled glass of oaky Australian Chardonnay then. That should calm her nerves.

REPLANTING GRAPE VARIETIES

Clonal experimentation, yeast selection, planting density, canopy management, minimal-intervention techniques, sustainability and organic practices, irrigation, and all other such measures will only work for so long. The inevitable next step will be to grow different grape varieties. And then, when that no longer works for some regions, to cease viticulture. It will all depend on how far we stretch the definition of wine—how wide its style parameters become, at which point the consumer no longer wants the resulting product, and at which point producing such a wine is no longer financially viable. One day, environmental manipulation will simply cost too much. So will the next step be to bypass the natural environment altogether? Is the future poly-tunnels?

Replanting could go a long way toward creating a sustained adaptation period throughout the world's winemaking regions. Grapes belong to either cool or warm climate groups, and within those grouping there is an order in which they achieve maturation for harvest: early-ripening or late-ripening.

Late-ripening varieties are considered best for helping regions adapt to climate change because, in general, the later grapes ripen, the better their sugar–acid balance. When grapes ripen too early, as they do in warm years, they fail to develop the acidity necessary for good balance. This means that many grapes that currently ripen in the optimal-harvest time of September are likely to become obsolete as temperatures increase. Ross Brown, CEO of Australia's Brown Wines tells us that "as the vineyards warm up a few degrees some of the varieties we are currently growing won't be viable in those vineyards in 10 to 15 years time."

Each grape variety has a specific taste and aroma potential, and a window of time during which this potential needs to be captured, or harvested. Using France as an example, the coolest climate grapes, grown in the northeast, are in the first period or *premiere époque:* Gewürztraminer, Pinot Noir, Pinot Meunier, Gamay, Chardonnay, Riesling, Sylvaner, and Pinot Gris. Those grapes in the warmer second period, grown in the west and southwest, are Sauvignon Blanc, Colombard, Chenin Blanc, Viognier, Semillon, Marsanne, Roussanne, Syrah, Cinsault, Merlot, Cabernet Franc, and Cabernet Sauvignon. And in the third period, the grapes best adapted to France's Mediterranean climate are grown: Clairette, Ugni Blanc, Grenache, Carignan, Cinsault, and Mourvèdre. In a more complete European profile, the range might be from Austria's Riesling to Portugal's Tourigo Naçional . . . two extremes. The farther north, the cooler and earlier-harvested/ripening the grapes; the farther south, the warmer and later-harvested the grapes. And it will be the late-ripening grapes that will help in adaptation.

Within each of these periods is, again, a harvest order based on maturation order, determined by climate. So if a winemaker in, say, Bordeaux grows Merlot, Cabernet Franc, and Cabernet Sauvignon, then the winemaker would harvest these grapes in that order. Merlot ripens first . . . Cabernet Sauvignon, last. At least, this is what usually happens. What is now happening is the warmer temperatures are blurring these maturation distinctions and the varieties are maturing "out of order," or at the same time. More confusingly, they're manifesting only sugar maturation as opposed to a full, phenolic maturation, as discussed earlier.

Grapevine Climate/Maturity Groupings

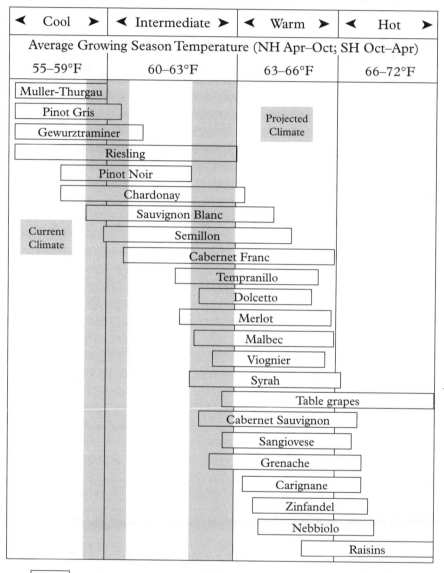

Length of rectangle indicates the estimated span of ripening for that varietal

"Climate Change and Wine: The Canary in the Coal Mine," by the Carbon Reduction Challenge. All research content provided by Gregory V. Jones, Department of Environmental Studies, University of Southern Oregon.

Note: This chart is of the climate-maturity groupings based on the current and 2050 projected average growing seasons in Oregon's Willamette Valley. It is a perfect snapshot of how eventually we will be attempting to grow our latest-ripening varieties in climates best suited to table grape or raisin production.

It will be the cooler regions that will enjoy the most sustained adaptation period. For example, Alsace, at the moment, grows the varieties of Riesling, Gewürztraminer, Pinot Gris (Tokay), Sylvaner, Pinot Blanc, Muscat d'Alsace, and Pinot Noir. These are six cool climate white grapes, all possessing high aromatic potential, and the red grape that requires the coolest climate. So as things warm up, Alsace has quite a lot of room in which to maneuver. Already its Pinot Noirs, which used to be slightly metallic rosés at best, are finding an *ampleur* and appeal akin to the lighter Burgundies. As some of the whites grow less acidic and move out of the fine wine category, the Pinot Noir could take their place. Then, eventually, the Pinot Noir might be nudged out of place by other red varieties, creating a gradual continuum. Replanting choices would slowly move down the "climate-category" scale . . .thereby forestalling any potential demise of viticulture in that region. That said, I am not able to envision Alsace ever not having the climate for wine production.

That is, if local appellation laws are allowed to change. This may prove to be the greatest obstacle to Europe's ability to adapt to climate change. For the New World wine countries, where there are no laws governing or limiting what is grown and where, replanting is second nature, a natural reflex, and a normal, current practice. In fact, this will be one of the single most relevant advantages the New World has in adapting to climate change. This freedom is priceless. But will the advantage of the replanting flexibility afforded by this liberal approach be enough of an advantage to counterbalance their handicap of possessing warmer climates? Most of the New World wine regions are already warmer than those in Europe to start with. Sorry to say it, but I thought Napa was too warm back in the 1990s, and suggested, quite sincerely, to Tim Mondavi that they think about port production. He was not amused. I have never enjoyed the New World wine style. In a recent tasting in London titled "California's Cool Climate Pinot Noirs," featuring wines from the "cooler" elevations and coastal regions, we were expected to treat clumsy, nondescript, and inelegant Pinots at 14 percent, 15 percent, and higher, seriously. When I pointed this out to one of the producers, he sheepishly admitted, "Well, they are cool for us." And none of the producers to whom I spoke had ever been to Burgundy. If you are going to put yourself forward as the world-class leader in Pinot-based fine wine production, should you not do your homework first?

EXHAUST THE RULE BOOK, THEN THROW IT OUT

Replanting need not mean panic in Europe, either. First of all, many of our best-loved regions have more legally permitted grapes than we know about with which they can begin. In Bordeaux, for example, we are familiar with the magic trilogy of Merlot, Cabernet Sauvignon, and Cabernet Franc and the role of Petit Verdot and Malbec, but there are fourteen red and white varieties permitted in Bordeaux under the INAO laws dating from 1935. For example, Gros Verdot, Malbec, and Carménère are also permitted although used sparingly. In Napa, ironically, these long-lost Bordeaux varieties are enjoying a renaissance.

Merlot was traditionally chosen in Bordeaux, specifically for its easy and early-ripening ability, a characteristic needed in the cool, wet, and tricky Bordeaux climate. But now Merlot is their biggest problem and is coming in with levels of alcohol at 15 percent. Producers are going back to Carménère. White varieties in Bordeaux are traditionally Sauvignon Blanc, Semillon, and Muscadelle, but Sauvignon Gris, Sauvignon Vert, Ugni Blanc, Colombard, Merlot Blanc, Ondenc, and Mauzac are all permitted, and all are later-ripening grapes that would adapt well to the changing climate.

I have spoken to producers all over Spain and Italy who are looking to the past for their future . . . researching and digging out older historical varieties that used to be gown in their regions. Spanish wine producer Bodegas Torres is leading, with the assistance of the Spanish government, a research project on wine's adaptation to climate change. Signor Torres Sr. has long been warning of the changes to come and been buying vineyard parcels in higher hillside parcels and cooler sites wherever he can. The company is also testing new rootstocks that resist drought or delay ripening to allow the aroma more time to mature. Their growers have recovered more than twenty forgotten Catalonian grape varieties that they believe will do better in high temperatures than the common varieties imported from northern Europe (aha, what have I been saying about keeping things indigenous?).

"The solutions are to be found in all the common experience of our forefathers," says Gérard Gauby, owner of the Domaine Gauby winery in the southern Roussillon region. Gauby advocates a return to ancient Mediterranean "Gobelet" vine training, in which the leaves form a protective umbrella over the grapes. He is also searching in North Africa for grape varieties like Grenache and Carignan, which have adapted to harsh climates. Likewise, in the southern town of Gaillac, Robert Plageoles of the Domaine des Très Cantous winery has started planting drought-resistant grapes like Ondenc and Verdanel, which he calls "the wines of the future." (Jeffery T. Iverson, "How Global Warming Could Change the Winemaking Map," December 2009).

Alastair Bland tackles this issue in his article "With Warming Climes, How Long Will a Bordeaux Be a Bordeaux?" "Changing grapevine varieties is absolutely an option, not in a short term, but by 2050 to 2070," Kees Van Leeuwen, of the Agricultural University of Bordeaux, is quoted as saying. "We are currently experimenting in our research center (the Institut des Sciences de la Vigne et du Vin) how later-ripening varieties behave in the Bordeaux climate. Changing French laws on regional grape use won't be easy, and could take as long as fifteen years. Other regulations are already breaking down—such as laws forbidding irrigation." He is referring to the 2006 INAO ruling. Bland points out that the practice is considered taboo by many, and some "wine critics feel it can result in watery-tasting wines." But he feels that "in exceptionally hot and dry conditions, watering vines can help increase the yield of fruit." And this is the point: As explained, this increase of yield comes with dilution. But this will not be enough to stop more regions following suit. They simply won't have the choice. And returning to the issue of salinity, we know that irrigation increases the soil's salinity, and salinity build-up in some southern Australian vineyards is causing winemakers to abandon them. The writing is on the wall, then.

Michel Chapoutier, a winemaker in the Rhône Valley, according to Bland, thinks that playing with the varieties currently permitted should provide enough leeway. "A Bordeaux will still be a Bordeaux, without a change in rules or a compromise in quality. Winemakers will merely have to adjust their blend ratios. Bordeaux [winemakers] will lower their amount of merlot and will raise their amount of petit verdot, while the Southern Rhône will lower their amount of syrah and raise their amounts of Grenache and Mourvèdre." I disagree; this may work for a very short while, but it is not the answer. Working within the appellation laws may at first prove effective. Adjusting their traditional "recipes" by changing the proportions of the grapes they use is only a stopgap, and one that is not afforded to those single-varietal vineyards, like Tuscany's Chianti Classico. When Sangiovese does not work anymore, what do they do? So the first place to start in many European regions is with the rule book—exhaust it. Then throw it out.

And speaking of Sangiovese, Bland presents the point of view of Lamberto Frescobaldi. The famous Tuscan says he "believes old traditions should not be allowed to hinder winemakers from adapting to climate change, even if it means permitting new grapes in their wines." But he doubts "that Tuscan winemakers will ever let Sangiovese, the main variety of the region, slip to the wayside."

He claims that "historically dominant wine regions will do everything they can to avoid giving up the grapes that have defined them for centuries." Begging the question, and echoing his title, is a Burgundy without a Pinot Noir or Chardonnay, a Burgundy? It may have to be. And speaking frankly, and not entirely politically correctly, these classic models are being deformed by heat anyway. Burgundy is no longer Burgundy, *with* the Pinot Noir and the Chardonnay grapes. And how many consumers really know that red Burgundy is Pinot Noir? That Rioja is Tempranillo? That Chablis is Chardonnay? Or that Barolo is Nebbiolo? And how many care? They just want good wine. And if the producers are savvy enough, they will find a way to make different wines under the commercially safe and stable umbrellas of their appellation "brands."

We have this perception that the traditional European plantings have been enshrined in ancient script. They'd like us to believe that. But they, too, have their dodgy pedigree. Bordeaux didn't really settle down to its current formula until this century, post-phylloxera, and I read that there was the odd Sauternes that had a bit of Riesling thrown in as late as the 1960s. There are producers everywhere quietly experimenting with other varieties. Oddly, I read that the Syndicat des Vins de Bordeaux asked permission from the INAO in 2009 to plant "non-native" grapes. Oddly, I say, because I cannot believe their motivation is so off target. A spokesperson from the syndicate was quoted as saying, "Bordeaux wines are blend-wines. We wish to test new varietals to know if they can enhance the complexity of our wines" (Steve Heimoff, "Bordeaux to Explore Non Bordeaux Varieties?" February 2009). Now, if this is really the reason that they wish to experiment, then I would have to be in agreement with their opposition. How would throwing in a bit of Syrah make the blend more complex? They have been accused of trying to "California-ize" the appellation and trying to cash in on more wine styles. But surely, they see this as an opportunity to explore varieties for future climate adaptation? Are they not noticing that their wines are deteriorating and do not understand why and so confusingly think adding more grapes as opposed to different grapes will make it better?

Or why not create some new varieties? Gaminot, Beaugaray, Picarlat, and Granita were developed by the INRA in Colmar and Sicarex-Beaujolais; they were crossings created at the request of wine producers in the Beaujolais Mâconnais, back in the 1970s. They are currently awaiting approval for entrance into the official national catalog. It takes time to create a grape variety. The life cycle of a vine is thirty to fifty years. It does not produce fruit for the first three years—so imagine the intricate combinations of crossings and then having to wait for them before you could make a wine from them to test and then allowing those wines to age so that they could be further tested . . . exhausting work. And to pass the rigorous criteria of the scientists, the proposed varieties would have to demonstrate that they preserve their varietal character from one generation to another, that they adapt easily to soil types while respecting a certain cluster density and berry size. The reason the producers wanted some new varieties is that they wanted grapes that would better withstand disease and so need fewer pesticides. But this was back in the '70s—is that still the most pressing problem our Burgundian producers face today? The Gaminot is a cross between Gamay and Pinot Noir and produces wines of great color that are more structured than Pinot Noir, but fruity like a Gamay; the Beaugaray, between Pinot Noir and Heroldrebe, is very aromatic; the Picarlat, between Gamay and Cabernet Sauvignon, is very structured, tannic, and deeply colored and fruity, best for blending; and the Granita, between the Auxerrois and Portuguese blue, which is, again, very aromatic. And they all apparently have completely new and original tastes and aromas. How exciting. I hope that we see more of this as crossings meant for climate change adaptation are attempted.

THE EUROPEAN PLANTING RIGHTS DEBATE: FINALLY, FREEDOM?

The Europeans are missing quite a few legislative tricks when it comes to plantation laws. They are fighting all the wrong fights. They are becoming their own worst enemy. For example, grape growers all over Europe are opposing a move by the European Parliament to relax planting limits in 2016. Currently there is a protective tool in place that limits vine plantings; it is due to expire then. This is a good thing. Right? But those who have vineyards in protected appellations do not want to see them expanded. They want to "protect" their product and feel that ending planting restrictions would "further weaken a sector that is already battling cheap imports, overproduction and declining demand." Translation: "But if there are more wines produced in my appellation, I will not be able to continue charging my exorbitant prices." This is incredibly shortsighted.

After a three-year battle, the European Commission renounced their decision to get rid of planting rights regulations and retain them until 2024, but French wine producers are calling for the European system of planting rights to be preserved until 2030. Proponents of the plan point out that wine is the only sector in European agriculture that is subject to planting restrictions, and that the European wine regions are the only restricted ones in the world. Countries within the EU will be allowed to increase their vineyard planting by up to 1 percent each year until 2030, when the system will be reviewed once more. Wahoo. The new scheme is so mired in legislative red tape that, frankly, it is ineffectual. It is obsessed with controlling production limits (prices and market control)—which it does not even achieve—and ignores the need to plant new varieties for either climate change adaptation or simply for the pure joy of experimentation, progress, or creative novelty. Think European: Change is bad! How does one tell them that they are risking all to protect a commodity that will no longer exist in 2030? By hanging on to these pseudo-protectionist policies, they are actually placing themselves in a far more vulnerable position. And who are the wine producers pushing for stagnation? When I am out in the field, the producers I meet want freedom to grow, change, and adapt. Do the two sides never convene? Apparently the act was pushed through by Member States that are nonproducing wine countries that wanted wine to be treated as any other agricultural crop.

But still, it is inconceivable that the European wine producers would try to restrict their flexibility when at the same time they are part of the international wine trade—the first industry sectors worldwide to agree to a consistent system to calculate carbon dioxide and greenhouse gas emissions, as devised by the International Organisation of Vine and Wine (OIV). Called the Greenhouse Gas Accounting Protocol, the system is set up to help companies assess the greenhouse gas emissions associated with their activities and also to offer guidance on the emissions associated with the vine and wine products. The UK's Wine and Spirit Trade Association backs it. So, while they are clearly up to speed on climate change (which is indeed what I am getting from my travels in the vineyards), some still have not connected the dots between climate change and the need for flexible replanting rights. Are they placing too much faith in the resilience of the *Vitis vinifera*? True, she is a feisty beast who has been adapting for hundreds of years to changing climates, and the hottest years in the classic European fine wine regions have often produced great vintages. They also have great faith in their own vineyard practices, experience, and savoir-faire, and rightly so. As Muriel Barthes, the technical director at the CIVB (Conseil interprofessional des vins de Bordeaux), points out: "In the past 30 years we have evidenced a marked rise in temperatures, but in the scale of 100 years, we know nothing, except for the new projection models." Perhaps their plan is to wait it out, hoping that the projections are not true and that another mini cycle of cooling will hit and save their vineyards for another few generations?

If it is commonly accepted that "dans cinquante ans, le Bordelais et d'autres régions viticoles auront des encépagements et des porte-greffes très différents"—in fifty years, Bordeaux and other regions will have very different rootstocks and grape varieties—then how do they suppose that they will get there if they fight any liberalized planting schemes? Serge Delrot, the director and a professor of the ISVV, continues to affirm that "if the droughts continue, it will be necessary to plant new, late-ripening grape varieties instead of those known to us today." He is far from alone; a fellow professor at Enita insists that there will be a time when wines have a different taste and typicity. And Gregory Jones asks, "If that ground is the best in the world for Pinot Noir, will it be the best for something else? If Burgundy warms to the point where Pinot Noir and Chablis are no longer the best grapes for that region, will people buy a new product? . . . At some point a grower in a region that has become too warm has to make a decision. It's like if you're selling widgets, and widgets don't sell anymore, what do you do? You have to come up with widgets 2.0."

The biggest problem with replanting, and one that some may find insurmountable, is that it is best for those regions that are starting new plantings or have the space for new plantings. If you are already established, if every parcel is planted, the life cycle of the vine makes this a very difficult process. Once you plant a plant, it is in the ground for fifty years. And it takes at least ten years to figure out it if you got it right—and by that time, the climate conditions for which you planted it will most likely have changed again.

The Germans are all up in arms. Like everyone else, they are petrified of the possibility of overproduction if planting laws are loosened or lifted. Steffen Schindler, marketing director for the German Wine Institute, says: "At last we are in a situation with no over-production but this would threaten that position." He also expresses concern that the decision could undermine Germany's recent success in raising its prices. With total exports in 2012 worth €321 million, the country's average price rose by 8.4 percent to reach €2.46 per liter. Again, they are thinking too short-term. They have hindered their own success in the global marketplace, not protected it.

In Spain, recent changes in appellation laws in Rioja have meant that we now have a "new" white Rioja made with newly approved grape varieties (as opposed to Viura). So it is finally happening. In 2007, the Rioja Regulatory Council approved six new white varietals: the local varieties of Tempranillo Blanco, Maturana Blanca, and Turruntés, and the international Chardonnay and Sauvignon Blanc and Rueda's Verdejo. The council agreed that while the local whites could stand alone, the international varietals could be no more than 49 percent of a white blend, with 51 percent reserved for local varieties. There has been a push to revive and revere the lost indigenous varieties of Rioja, especially the reds, like Maturana Tinta. Again reviving the argument that indigenous is best: The Maturana Tinta, with its late budburst, early ripening, high acidity, and medium alcohol content, is far better suited to the warming climate.

But again, the push for these changes may not be for the reasons you might think. The main reason is Rioja's desire to produce more individual and *terroir*-driven wines and to move away from the international obsession with the Cabernet Sauvignon and Merlot. Winemakers in Rioja do not see the point in trying to compete with the rest of the world to make the same wine when they have such individual and unique varieties at their feet. This is to be applauded. Apparently Rioja winemakers have also been frustrated with the mediocre results of the Viura for years, partly because Rioja whites are made the same way as their reds—oaked and heavy. They are also frustrated that whites from other Spanish regions, Rías Baixas and Rueda, are having a greater commercial success than theirs. But when I spoke to Viura producers, they all told me that the heat means Viura wines are becoming increasingly insipid, especially in Rioja Baja, the southernmost and warmest of the La Rioja subregions, where the climate is more Mediterranean than Continental and the red wines can reach 18 percent alcohol.

Although this is not the reason being given for the stylistic shift. Again, several producers told me that either people are still not connecting the dots between rising temperatures and their increasingly boring wines, or, if they have, it is not a subject they wish to highlight, for fear of alarming the consumer and upsetting the market. There is also the same fear that the relaxed laws will be exploited by some producers who intend to make more international wines and in great quantities. The new planting laws can be used to pursue two different market strategies/goals. But this has always been the case, everywhere in the world (and in all industries). Those who wish to produce mass-market wines will never be restrained by legislative tools; they will always find a way around them. While those whose intention is to produce indigenous, traditional wines must not have their efforts confined nor stunted by these same tools.

Not surprisingly, England is one of the few nations that support the EU's "liberal" planting move. "We should have the freedom to plant where we want," says Julia Trustram-Eve of English Wine Producers. "Planting restrictions have held Europe back, allowing some areas to continue to produce poor-quality wine and falsely inflate the value of land." Perhaps the change is simply too frightening to face.

The restriction in surface area to a rise of 1 percent each year (if you are in a certain area, if you promise to grub up an equivalent amount of vines first, if you only plant varieties that were already authorized, if you belong to a Member State that applied before 2007, or if you name your firstborn either Adelbert or Anique and promise to serve them *frites* at every meal) is a frantic and futile attempt to avoid overproduction. Smart—limit production some more when your yields have already been decreased between 20 and 40 percent due to recent harrowingly disastrous harvests. An EU representative complains that "the wines of Bordeaux and Cognac endured a terrible crisis due to a large increase in the number of vineyards during the 2000s." He further argues that sales of Australian wines were booming in the 1990s and became "the international model of success in the winemaking world, and particularly in exports." The positive reaction to this export success led to a slew of new vineyards in the absence of any laws limiting quantities. Australia, he says, now faces "uprooting, bankruptcy, non-harvesting, sequestration, a rapid drop in land values, the purchase of assets by foreign investors." Yes, but not due entirely to overproduction, but to the cost overhead needed to force wine production in a hostile environment and the oversalination of the soils. Even if they had stemmed their production, these problems would not have gone away.

If I were a European wine producer, I would be furious and start looking for loopholes. In the EU's working document of April 2012, Rule 1 under planting rights granted from a Reserve, states that "the location, the varieties and the cultivation techniques used guarantee that the subsequent production is adapted to market demand." It explains that these qualitative criteria are not determined, which gives a Member State room for interpretation on a regional level. These criteria also allow "taking into account territorial considerations. For example, it may be decided that no planting rights are granted to certain regions of a Member State, to certain types of soil-climate conditions or for the establishment of vineyards which so not respect certain technical specifications, as long as this is justified with—lack of adaptation to market demand." So if my Meursault was so flabby and sweet that it no longer fit the classic model of Meursault and I could not sell it as such, does that not mean that my product no longer is able to adapt to "market demand"? And doesn't that mean that I should be able to grow a grape in my "soil-climate conditions" that would adapt to market demand? Perhaps this is how they are secretly hoping that these instruments will be applied. They like a good haggle and a bit of jumping through hoops. And what about the good old "force majeure" card? I'd petition the INAO for a *demande d'autorisation de plantation de vignes* based on a few *intempéries graves* (severe weather conditions). Or use the rule where you are allowed "experimental" plantings as certain Bordeaux châteaux are doing, such as Château Cheval-Blanc or Larose-Trintodon in the Médoc. I know that the Cathiards, owners of the Château Smith-Haut-Lafitte, have an entire private island in the Gironde dedicated to their clonal and varietal experimentations.

MUTINY IS THE ANSWER

Another fight that seems to be being waged for the wrong reasons is the one waging in Sancerre. When the INAO (l'Institut national de l'origine et de la qualité) announced that it was closing its office in Sancerre and centralizing it in Tours, 210 kilometers away, the producers, represented as the Union Viticole Sancerroise (UVS), were angered at the fact that they would still be paying their fees to the INAO but receiving a reduced service. They are now considering leaving the INAO and creating a "Sancerre" trademark and are seeking legal advice. Critics of the move take a conservative stance and suggest that such a move would destabilize the entire appellation system. But again, so what? It is being destabilized as we speak, anyway. I think that the producers should take back their brand and do with it what they like. They all seem to be so protective and overworried about production levels, but in the grand scheme of things Sancerre is minuscule and its sheer physical restrictions should serve as sufficient protection. Leaving the INAO is technically possible and if done properly would not endanger the cachet of its AOC value. It does not need to be called AOC, it needs to be called Sancerre and it needs to be good. Whether that means it is Sauvignon Blanc from Sancerre or one day, perhaps a warmer climate varietal such as Viognier from Sancerre. The market forces that created Sancerre will sustain Sancerre. Further, there is so much diluted, flabby, and overly worked Sancerres being produced anyway. The use of heavily flavored selected yeasts is so marked: Sancerre *à la fraise*, *à l'ananas*, or *à la banane*, anyone? So while they are all frightened of leaving the false safety umbrella of the INAO, and of loosening their grip on their monopoly of mediocrity, they are risking giving up a chance at future freedom.

Another point: The planting restrictions only apply to those who wish to comply to the system of controlled designation of origin (AOC). So in theory, you could boycott the planting rules but lose your appellation and become a *vin de table*. But then you risk not selling your wines at the same price anymore. Unless everyone followed suit and looked to Italy's Super Tuscans for inspiration. They broke away from the sloppy and illogical varietal Chianti laws and created their own version, treated it like a brand, and were able to charge three times more than the regulated appellation wines.

POLY-TUNNELS AND GREENHOUSES: "FERTILE DESERTS"

Moving down the sliding scale of adaptation options . . . we must consider the possibility of indoor cultivation. Perhaps before this, some may explore netting and screening. Nature had intended the vine's foliage to handle the job, but this self-regulating system has broken down in arid climates. I know of a Riesling producer on the slopes of the Mosel who started screening a particular parcel from the sun during a week or so here and there. He began this over five years ago. He found he was using the screening more and more each growing season, until finally he grubbed the Riesling and planted a few red varieties, such as Syrah, to experiment, tired of watching his Riesling become parched and sunburned.

Netting and screening are expensive and impractical solutions for any large-scale grape growing, obviously. And so is indoor cultivation, really. Any system would have to be ecologically efficient to engine or the cost—to both the producer and the environment—is prohibitive. But some argue that if there is new technology out there being used for other large-scale crop production, then wine grapes should also be considered.

One approach is powering greenhouses geothermally. Geothermal heat for greenhouses usually relies on soil and water belowground containing a vast reservoir of thermal energy. Geothermal heating systems recover this energy and convert it to heat that can be utilized in greenhouses and other buildings.

Greenhouse consultants Hubert Timmenga of Vancouver and Peter Klapwijk of the Netherlands have brought Innogrow's patented GeslotenKas closed greenhouse system to British Columbia. This very new technology is still being tested, but promises to reduce input costs, improve yields (in the cool-climate context), and reduce disease. Closed systems collect and store energy from the sun without releasing it into the environment. According to Innogrow, they provide "climate control" instead of "climate management" through geothermal heating and cooling. Geothermal heating/cooling is becoming common in homes. A closed-loop version is being developed for the NASA Mars program, and Agriculture and Agri-food Canada is using an open-loop version at the Pacific Agriculture Research Centre in Agassiz. The first commercial-scale Innogrow system was built for the 5.4-hectare Themato greenhouse in Holland in 2004. The costs of the electrical power to engine this system are still considerable.

There is a company, Sundrop Farms, that has developed a proprietary system to grow food in some of the world's driest regions using abundant and renewable resources—sunlight and seawater—to create "fertile deserts." Sundrop explains that agriculture can be supplied by two kinds of water: water that comes in the form of rain, "green," and water that is stored in rivers, lakes, glaciers, reservoirs, and underground aquifers, or "blue."

> While rain fed water is essentially free for farmers, it is finite. Once the last rain-fed hectare of land has been put into production this water resource is exhausted. As such, green water has to be augmented by blue water, which is what irrigation during the last centuries has been all about. Globally, irrigation now accounts for 70% of the 3,240 cubic kilometres of water withdrawn for human use. Tapping into blue water resources and expanding irrigation lands has allowed us to feed our growing population until now. However, unlike green water which is self-regulating, i.e. once you use up all the rain water there is no more to apply to a field, blue water resources have been exploited faster than they can be replenished . . .

The Sundrop Farms System harnesses the sun's energy to desalinate seawater to produce fresh water for irrigation, electricity, and heating and cooling energy. In 2010, the firm began operating the world's first commercial Sundrop Farm in South Australia, located at the top of the Spencer Gulf, north of Goyder's line, near Port Augusta. Given the lack of fresh water, degraded pastureland, and harsh climates, traditional agriculture struggles in this area. Sundrop technologies allow produce to be grown from Southern Ocean seawater and sunlight. A twenty-hectare greenhouse produces over fifteen thousand tons of tomatoes annually, a planned eight-hectare expansion is expected to produce 2.8 million kilograms of tomatoes and 1.2 million kg of peppers year annum, while saving the equivalent of about 4.6 million barrels of oil (equivalence) and 280 million liters of fresh water, compared with a standard greenhouse in a similar location.

There are many similar new technologies being investigated for greenhouses. Would they ever be envisaged for use in wine grape growing? If that doesn't work, the next step is the cessation of viticulture, or the replanting the vineyards in new regions . . . and this is where the wine remapping really begins.

4

WHAT REMAPPING MAY LOOK LIKE: TO CEASE, ADAPT, IMPROVE, OR EMERGE?

The eeriest event has just occurred. As I was rewriting this chapter for the umpteenth time, I had decided to begin it with some of the climate data that Francesco Valentini gave to me during a visit to his home last summer in Loreto Aprutino, Abruzzo. This is the most elusive, enigmatic, and quality-obsessed family to exist in winemaking history, and they have been making wine since the 1600s. It was such a pleasant surprise to discover our mutual passion for climate change discussion, and I was thrilled as he thrust a bundle of family records into my hands. They had been monitoring the increasing temperature rises and the resulting earlier harvest dates since the 1800s.

Lost in a reverie evoking that sun-drenched afternoon, heavy with an oppressive heat unlike any other, my thoughts wandered from the page. The Valentini home is down a narrow medieval alleyway, and we had to abandon our vehicle and walk. Hot, molten-black paving stones played havoc with my impractically strappy sandals. The midday heat burned into the back of my neck, and even the slightest movement triggered discomfort. One of my colleagues muttered under his labored breath that even his fingernails were sweating. Though we were surrounded by a cacophony of vibrant flowers and miles of olive branches that clung to every doorway and windowsill, the heat muted their odors; we could only imagine their perfumes. We were half an hour inland from Pescara and the coast—if only the merest whiff of a salty breeze would waft toward us and break the solid wall of Hot. Relief came as we were ushered into the shaded courtyard of their villa, and then into a dark, cool library. It did not take very long before I was ensconced in a silk settee and we were offered their 1976 Trebbiano. Words fail to describe the effect as this clear golden nectar trickled and danced a jig down my parched throat. It was as fresh, lively, and crisp as the year it was made . . . its subtle personality and charm erasing all woes.

So imagine my disbelief when this reverie was interrupted by a call from a colleague who tells me that a blizzard swept through Abruzzo, destroying vineyards with its freezing temperatures and Arctic winds. Francesco Valentini was interviewed by *Corriere della Sera* journalist Luciano Ferraro. Valentini explains:

> In the provinces of Pescara, Chieti, and Teramo, there was a first wave of bad weather two weeks ago (mid-November), with 500 millimeters of rainfall in just a few short hours. Then, last Tuesday, there was a blizzard, with gusts up to 100 kilometers per hour. It wreaked havoc in our vineyards. We're still assessing the damage but I believe Abruzzo had up 2,700 hectares under vine that were leveled. Half of the vines were damaged, unfortunately in the best zones for both Trebbiano and Montepulciano. Precious vines that were at least a half-century old. At this point, we'll have to survey each plant to determine which can be saved and which will be thrown away. The storm brought tremendous damage. And many other wine and olive oil producers are in the same situation.

Apparently, because many vineyards in Abruzzo use the Pergola trellising system, the weight of the snow on the canopies caused them to collapse. A friend has also spoken to him, and he says that despite about 50 percent of his vines being damaged, he plans to produce wine this year; he will know better in the spring, he adds, what the real damage is. It is difficult to reconcile the recent images of the snow-covered vineyards with the memories of my summer visit.

And that 1976 Trebbiano we enjoyed? It was harvested on October 19 of that year. In the past decade, they have consistently harvested the Trebbiano by mid- to late August. And to just complete my thoughts on the Valentini family, here is a quote from two of their many admirers, Sheldon and Pauline Wasserman (*Italy's Noble Red Wines*, 1991) regarding Francesco's eccentric late father, Edoardo:

> "His first selection is in the vineyards. If it is a rainy, but not too rainy year, he selects the fruit from the vineyards with a southern exposure; in drier years he chooses grapes from vines facing more northerly. He selects the part of the vineyard least affected by the weather and

then selects the best bunches. The rest of the grapes are sold. In the years when he produces wine to bottle, about five percent of his best grapes are turned into wine; the rest of the fruit is sold. At most he makes 50,000 bottles of wine a year. Generally he produces much less. Average production, in the years that he produces, is more like 5,500 bottles of red and 22,000 of white. He selects from the wine he produces the best to bottle and rejects the rest, usually most of the production. Would that more producers had his integrity.

Selected Latitudes (Northern Hemisphere)

Wine Region	Latitude ° North
Södermanland County, Sweden	58.58
Gotland Island, Sweden	57.50
Scania, Sweden	55.90
Jutland, Denmark	55.62
Mohe County, China	52.10
Kent & Sussex, UK	51.00
Mosel Valley	50.05
Pfalz	49.4
Okanagan Valley, British Columbia	49–50
Champagne	49
Finger Lakes	48.5
Alsace	48.5
Tokay	48.1
Thurgau, Switzerland (NE)	47.50
Chablis	47.4
Sancerre	47.33
Loire Valley	47.17
Austria	47
Côtes du Beaune	47
Lake Geneva	46.46
Alto Adige	46.38
Columbia Valley, Washington	46–47
Beaujolais	45.8
Veneto	45.73
Cognac	45.62
Piedmont	45.25
Bordeaux	45–46

(continued)

Wine Region	Latitude ° North
Nova Scotia	45
Willamette Valley	44.9
St. Emilion	44.89
Traverse City, N. Michigan	44.70
Rhône Valley (Châteauneuf du Pape)	44.13
Chianti	43.52
Languedoc	43.43
Provence	43.43
Tuscany	43.35
Niagara Peninsula, Ontario	43.00
Roussillon	42.03
Penedes	41.38
Douro	41.23
North Fork, Long Island	41.05
Apulia	41.01
Alentejo	40.31
La Rioja	39.57
Ribera	38.73
Napa Valley	38.50
Sonoma Valley	38.17
Nemea, Peloponnese	37.8
Etna, Sicily	37.75
Loudoun County, Virginia	37.50
Ningxia Province, NW China	36.60
Moscato di Noto, Sicily	36.44
Santorini, Greece	36.41
Mount Fuji, Japan	35.35
Sonoita, Arizona	31.6

THE WINE WORLD REMAPPED:
A SAMPLING OF PROJECTIONS

Using a consensus of all the climate data, reports, and opinions I have re-searched, and armed with my 1906 edition of *Harmsworth Atlas and Gazetteer*, I have attempted to superimpose the world's vineyard maps onto the pro-jected shifts in latitude north for the Northern Hemisphere and south for the Southern Hemisphere. For example, Germany's coordinates are 10W–20E,

40–60 N. We know that *Vitis vinifera* is most suited to climates within the 30th and 50th parallels in both hemispheres. Assuming that 60° is soon to be the new 50°, and the Mosel Valley sits at 50.4°, we can look forward to German vineyards on the Baltic Coast (54–55°) before Germany is no longer able to expand wine production northward. And conversely, with 40° being the new 30°, and with most of southeastern and southwestern Australia being between 30° and 35°, these regions are falling out of the suitability zone. I tend to move back and forth between the general and the specific. Please forgive me if it seems repetitious, but it felt relevant to do so, as this represents the nature of the provided data. There is often a greater amount of consensus regarding a more general area and less consensus within the smaller regions, and there has been very little work conducted on specific microclimates. There will be errors and omissions here; I do not have a crystal ball. But I was unable to resist a go at this exercise, albeit rudimentarily. There are many more factors at play than latitude.

Apart from the use of latitudinal positioning, another approach is that taken by PNAS: the measurement of vineyard area suitable for viticulture due to the impact of water stress. The two approaches merge and come to similar conclusions. They would, because they are, in fact, the same measurement: Higher temperatures equal less water. For example, in the case of Chile, PNAS states that it will experience a 47 percent decline in viticultural area due to the impact of freshwater conservation and is currently experiencing a 94.6 percent level of water stress with predicted precipitation trend of a 15.5 percent average decrease to the period of 2050–2080 (the general agreed baseline measurement is 1950–2000). Chile's classic viticultural areas fall between 30° and 36° and are moving ever southward. Its most recent migrant is Casa Silva, planting farther south than Bio Bio, on the northern shores of the Lake Ranco, at 40° 1451" S. When dealing with altitudes within specific wine regions, I have considered replanting to higher sites, where possible, as an adaptation measure: Regions will attempt to plant higher before they consider planting farther south or north, assuming their land surface would permit such a move. Clearly, altitudinal parameters will be exhausted before latitudinal shifts will be, and when these latitudinal shifts are made to cooler regions, they may be at lower altitudes. Until, that is, that new region becomes too warm, and plantings are then again moved to higher altitudes before another shift of latitude is taken. There are regions of greater altitude but lower latitudes that will be as cool as or cooler than lower-altitude regions at higher latititudes. For example: Mount Fuji has a latitude of 35.5° but a Köppen classification of ET (tundra), while Mohe County, China, is at latitude 52.10°, but has a warmer Köppen classification of Dwc (subarctic).

There is also the variable of inland versus coastal within the same latitude: The former will be warmer than the latter.

For example, California's Central Coast spans an area from Monterey Bay to Point Mugu and comprises Santa Cruz County, San Benito County, Monterey County, San Luis Obispo County, Santa Barbara County, and Ventura County. Running parallel, but inland to the east, lies the Central Valley, roughly from Stockton to Bakersfield. The center point of each, respectively, is Monterey and Fresno. Monterey has a latitude of 36.6° and Fresno, of 36.7°, so the same. Yet Monterey has a cool summer Mediterranean climate (Köppen classification Csb) while Fresno has a semi-arid Mediterranean climate (Köppen classification Bsh), and their average summer temperatures are degrees apart.

Imagine a sort of slow belly dance of beaded climate bands undulating both up and down and in and out, chiming in at different times with different voices, but still in tune, like in a round. Voices singing the same melody but each voice beginning at a different time so that different parts of the melody coincide in different voices, but still fit harmoniously. As with a round, there is only the one line of melody that needs to be learned by heart and by all. This is our challenge.

Remapping the World's Vineyards

Cease or Decrease

Yields in these regions will significantly decrease due to increased drought, increased humidity, or increased precipitation, or will cease viticulture after having struggled through a stage of adaptation measures. These are regions, generally, that are already suffering crop loss from heat and drought and are already heavily irrigating.

EXAMPLES:

Southern, Inland Mediterranean

Andalucía, Spain
Douro Superior, Douro Valley, Portugal
Languedoc-Roussillon, France
Northern Africa
Rioja Baja, La Rioja, Spain
Sicily (not Etna)
Southern Italy

Western North America
Central Valley, California
Napa and Sonoma Valleys
Sonoita, Arizona

Inland South Africa

South America
Maipo Valley, Chile (and northern Chile)
West-central & and Northern Argentina

Southeast & southwest inland regions, Australia
Adapt
These are regions with the ability to remain in the same area and
continue a viable viticulture assuming that adaptation is achieved by
moving to higher altitudes or to coastal regions, implementing irrigation, or making changes in grape varieties and/or winemaking styles.
Most of the Old World classic wine regions are in this category. There
has to be a concerted effort to stop making overly extracted wines first
so as to strip away this variable and observe the true effects of climate
on the wines—and then react. These regions will face an identity crisis
as their wines change style; resisting the temptation to overproduce,
and using their appellation as a "brand" tool, effectively, will be vital.
They will survive and improve by replanting if they have room for
sliding down the warm climate variety scale. An example: Alsace's
white grapes being replaced by more suitable red varieties.

EXAMPLES:
Alsace, France
Baixo Corgo and Cima Corgo, Douro, Portugal
Bordeaux, France
Catalonia, Spain
Central and Southern Serbia
Chablis
Coastal Southwestern Australia
Columbia Valley, Washington
Coonawarra, Coastal Southeastern Australia
Koper, Istrian Peninsula, Slovenia
Loire Valley
Lower Sava Valley, Slovenia

Odobesti, Panciu, and Nicoresti, Romania
Pacific Northwest, USA
Piedmont
Rhône Valley
Rioja Alavesa and Rioja Alta, La Rioja, Spain
Tuscany
Veneto
Vinho Verde, Portugal

Improve
These that currently have an established viticulture but are at the cooler outer margins of production and will ameliorate due to the changes, eventually, and see a more reliable output (not necessarily a "better quality"), and/or increase in suitable viticultural area. This is the most likely category in which to site future fine wine. These regions are both improving and emerging, really, as they expand. Their progress may be impeded by more extreme smaller wet-weather patterns, but the overall trend will lead to better ripening. This category includes many of the northern white wine producing regions (like the Rhine) that will move into heavier white wine production and into increased red production—so *improve* is an unknown term in this case. An example would be Alsace's improving Pinot Noir. Also included in this category may be those undeveloped regions that have previously taken a small role on the world wine stage but now may find that climate change gives them an opportunity to step up their game.

EXAMPLES:
Alto Adige, Italy
Balti, Moldova
Bio Bio and Andes foothills, Chile
Brda, the Littoral, Slovenia
Central Otago, New Zealand
Drave Valley, Slovenia
Finger Lakes, New York
Great Dividing Range, Australia
Hudson River Valley, New York
Lake Geneva, Switzerland
Luxembourg
Malokarpatská, Lesser Carpathia, Slovakia
Mosel Valley, Germany

Northern Austria
Northwestern Spain
Northwestern USA
Nova Scotia
Okanagan Valley, BC
Rhine Valley, Germany
Snake River Valley, Idaho
Southern England
Southern New Zealand
Subotica-Horgoš, northern Vojvodina, Serbia
Vipava Valley, the Littoral, Slovenia
Willamette Valley, Oregon

Emerge
The following regions already see some fledgling viticultural efforts, or will emerge as entirely new production zones in areas where there currently does not exist any viticulture.

EXAMPLES:
Balmaceda, Chile
Bohemia Mountain, the Cascades, Oregon
British Columbia
Canadian Rocky Mountains
Flagstaff, Arizona
Gotland Island, Sweden
Hobart, Tasmania
Jutland, Denmark
Mohe County, China
Mount Fuji, Japan
Mount Washington, New Hampshire
Northern England, Scotland, and Ireland
Northern Europe
Northern France (Brittany & Normandy)
Quebec
Regions in northern China
Scania, Sweden
Södermanland County, Sweden
Southern Patagonia, Chile, and Argentina
Thurgau, Switzerland
Yellowstone Park, Wyoming
Yukon

THE DISSENTING VIEW

One of the prime sources in this debate is the April 2013 PNAS study to which I have referred on numerous occasions in this work—*Climate Change, Wine and Conservation*. This is the study in which the authors (Hannah et al.) write of the possibility that not only will climate change render large swaths of the Mediterranean and other vineyards as unviable, by as much as 25 to 73 percent by 2050, but that it will go on to shift animal species as well, as our changed land use could result in habitat loss:

> Climate change has the potential to drive changes in viticulture that will impact Mediterranean ecosystems and to threaten native habitats in areas of expanding suitability. Redistribution in wine production may occur within continents, moving from declining traditional wine-growing regions to areas of novel suitability. The Mediterranean is expected to increase temperatures along with less precipitation over the coming decades, meaning viticulturists in the region will need to change the varieties of grapes they're growing, use alternative irrigation systems or stop growing wine grapes altogether.

In a more recent PNAS paper, written as a response to Hannah et al., *Why Climate Change Will Not Dramatically Decrease Viticultural Suitability in Main Wine-Producing Areas by 2050*, van Leeuwen et al. suggest that Hannah's conclusions were alarmist. Van Leeuwen et al. agree that the expansion of viticulture into new areas "can" lead to a decrease in biodiversity and that an increase in water use for irrigation "might" lead to major freshwater impacts, but disagree "with the alarming statement that suitability for winegrowing of main wine-producing areas world-wide will dramatically decrease over the next 40 years." They assert that Hannah's conclusion that most of the present wine-growing regions will become "unsuitable for viticulture" is "erroneous" and identify three flaws in Hannah's work (all "methodological"): the misuse of bibliographical data to compute suitability index, the "inadequacy of the monthly time step in the suitability approach," and "an underestimation of adaptation of viticulture to warmer conditions." They have a problem with Hannah's grapevine groupings as defined by Jones, in his *Climate Change and Global Wine Quality* paper. I have included his table on the following page.

This table is from Gregory V. Jones's Paper 4 from the "*Terroir*, Geology and Wine: A Tribute to Simon J. Haynes" session held at the Geological Society of American Annual Meeting, Seattle, Washington, November 2, 2003.

Region	Growing Season (a) Tavg (°C)	Climate Maturity Grouping (b)
Mosel Valley	13.0	Cool
Alsace	13.1	
Champagne	14.5	
Rhine Valley	14.9	
N. Oregon	15.2	Intermediate
Loire Valley	15.3	
Burgundy—Côte	15.3	
Burgundy—Beaujolais	15.8	
Chile	16.3	
E. Washington	16.5	
Bordeaux	16.5	
C. Washington	16.6	
Rioja	16.7	
S. Oregon	16.9	
C. California	17.0	Warm
South Africa	17.1	
N. California	17.4	
N. Rhône Valley	17.6	
N. Portugal	17.7	
Barolo (Piedmont)	17.8	
S. Rhône Valley	18.2	
Margaret River	18.6	
Chianti (Tuscany)	18.8	
Hunter Valley	19.8	Hot
Barossa Valley	19.9	
S. Portugal	20.3	
S. California	20.4	

Mr. Jones writes: Wine region average growing season temperatures as analyzed by Jones et al. (2005) sorted into their respective climate maturity groupings as depicted in figure 2 (see page 102). Note that the growing season average temperatures depicted here are derived from a 0.5° × 0.5° grid and not from any one station.

a. The growing season is Apr–Oct in the Northern Hemisphere and Oct–Apr in the Southern.

b. The climate maturity groupings are based upon the average growing season temperatures and the ability to ripen a given variety.

Hannah is not saying that all wine-growing regions will become unsuitable, but rather that the high-quality regions will have their quality compromised and that some regions that are already too hot, and are presently producing the world's bulk wines, will suffer a demise, one way or another. This is how I interpreted his work, and I hope that I have done so correctly, and it seems to fall into line with Dr. Gregory Jones's work.

Again, it is important to acknowledge that climate change is not the sole factor in the vintage quality ratings. We know that significant improvements in winemaking practices, and significant preferences in winemaking trends, combined with the warming climate to create these higher ratings, but decreased year-to-year variations. We also know that weather and climate are two different things and that they both affect the grape's growing season in different ways. The best way to explain it might be to say that our weather patterns are a subset of the larger climatic changes, but they are both working simultaneously, and often are at odds with one another. This must mean, then, that the effects of climate change will not be the same in all the regions, or for even all the grape varieties in a given region, even if they are all experiencing the same warmer weather.

In their *Climate Change and Global Wine Quality* (2005), Jones et al. state that "from 1950 to 1999 the majority of the world's highest quality wine-producing regions experienced growing season warming trends . . . Currently, many European regions appear to be at or near their optimum growing season temperatures, while the relationships are less defined in the New World viticulture regions. For future climates, model output for global wine producing regions predicts an average warming of 2°C in the next 50 years." Again, what they are explaining is that the known zones for quality wines are increasingly becoming unsuitable for their current varieties and that these zones are shifting—varieties are being "pushed out" of their previous, classic climatic zones.

Van Leeuwen et al. argue that the Jones groupings were "constructed from empirical observations collected in premium wine-growing areas and not based on grapevine physiological modelling" and argue "that it is very difficult to establish precise upper limits by variety for growing high-quality wines and that those given are underestimated." To evidence their argument, they

compared average growing season temperature (AvGST) from 1971 to 1999 and from 2000 to 2012 for three major wine-growing regions: Rheingau (Germany), Burgundy (France), and Rhône Valley (France). Burgundy continues to produce great wines with Pinot Noir since 2000, although AvGST is already above the upper temperature limit cited (See Table). The same is true for Rheingau with Pinot Gris and the Rhône Valley with Syrah. High-quality viticulture is sustained in these regions despite increased temperatures and dry farming, because of both the evolution of consumer's preferences and implementation of adaptive strategies by growers."

I disagree.

OPTIMAL AVERAGE GROWING SEASON TEMPERATURES FOR 21 COMMON VARIETALS OF *VITIS VINIFERA*

Variety	Low temperature, °C	High temperature, °C
Muller-Thurgau	13.1	15.05
Pinot Gris	13.1	15.3
Gewürztraminer	13.1	15.65
Pinot Noir	14	16.2
Chardonnay	14.05	17.15
Sauvignon Blanc	14.65	17.7
Riesling	13.2	17.1
Semillon	14.9	18.15
Cabernet Franc	15.35	18.9
Tempranillo	15.9	18.6
Dolcetto	16.4	18.55
Merlot	16	18.8
Malbec	16.25	18.95
Viognier	16.6	18.8
Syrah	16.15	19.15
Cab Sauvignon	16.4	19.85
Sangiovese	16.9	19.5
Grenache	16.6	20.1
Carignan	17.15	20.2
Zinfandel	17.5	20.5
Nebbiolo	17.6	20.9
All varieties	13.1	20.9

Hannah et al., *Climate Change, Wine and Conservation*, PNAS.

Van Leeuwen also uses as one of their supporting references the work of John Gladstone, who supports their thesis that the effects of climate change are being over-exaggerated. Gladstone, in his *Wine, Terroir and Climate Change*, summarizes the implications of climate change for vine *terroir*s:

Recent concerns for future viticulture and its *terroir*s have focused almost entirely on climate change and, in particular, greenhouse warming. That would be justified were the more alarmist of greenhouse predictions to prove true. Substantial migrations of viticultural regions would then be needed, or of grape varieties and wine styles within them. But critical examination shows the claims of greenhouse warming to have been poorly based and almost certainly much exaggerated. It remains a reasonable expectation that some greenhouse warming will occur through the 21st century, but hardly enough to raise temperatures materially above those of recent decades until after mid-century.

MAY, PERHAPS MAYBE, POSSIBLY MIGHT

Gladstone is skeptical but seems to want to hedge his bets as he concedes, just in case, that "any warming will certainly allow the spread of viticulture pole wards and to higher altitudes, and such vineyards may be needed to maintain some cool-climate wine styles in their purest forms. These vines will, however, be least able to exploit the higher CO_2 available to them. Hot and dry inland viticultural areas, though suffering disproportionate heating and drying, may compensate through the vines' greater water use efficiency and heat tolerance. But effect on wine quality will probably be adverse."

One is given the impression throughout all this research that there is not too great a divide among members of the scientific community. They agree upon similar future scenarios if temperatures continue to increase at the rate that has been forecast. The debate seems to be centered on this rate, this looming time line. Disappointingly, they also seem to be squabbling over the concept and definition of *terroir*, the effects of irrigation, the optimum temperatures for grape varieties, and what constitutes a good wine. These are concepts long since accepted and understood by nonscientists. And those climate change scientists who are also viticulturists, those who "get it," are exhausting themselves trying to convince their less enlightened colleagues. When Van Leeuwen et al. write that they believe that both Burgundy and the Rhône Valley are still producing "great wines" with Pinot Noir and Syrah, I have to question their wine-tasting experience. I was amused to come across one study by a group of Canadian geoscientists who were tackling yet another paper on the link between *terroir* and fine wine, and who did so by repeated "evaluations of samples from various studied vineyards." If you are coming at the subject of climate change and wine from a climate change perspective and have to bone up on your wine knowledge, your learning curve will be too great. It is far better to approach the topic possessing strong wine knowledge and vast tasting experience (as do Jones, and Smart, for example).

Science is a form of human activity through the pursuit of which humankind attempts to acquire a fuller and more accurate knowledge and understanding of nature—past, present, and future—and an increasing capacity to adapt itself to, and to change, its environment, and to modify its own characteristics. The most crucial principle in scientific research has always been historical observation based on practical experiences, and climatologists have always relied upon the tools of observation to reconstruct the history of weather: ancient texts and depictions, old journals, newspapers, shipping logs and forecasts, farming records. Humans have been recording the changes in their physical environment since they had to create fire to warm their caves. Hence it is the recordings of the world's grape growers, this precious primary source of practical experiences, that will prove to be the most reliable and richest of sources.

Most important, I have listened to the voices of those men and women who are not scientists, but experienced winemakers . . . the grape farmers. Those who have meticulously recorded every nuance of their land and the effect weather is having on it. They alone best understand the changes and challenges before them, and they alone are the most qualified to remap our future vineyards.

From Austria to Australia, the conversations are rife with what-ifs. The principal and universally accepted climate projection models are looking thirty to fifty years ahead. The Köppen-Geiger Climate Classification, using the IPCC projected data, has provided a model for the period 2076–2100.

Using these maps, it is easy to see how current regions will be shifted in and out of their current climate classifications. The Intergovernmental Panel on Climate Change predicts that the earth's temperatures could rise by as much as 11°F (6°C) by 2100 if nothing is done to combat climate change. "While 2–3° C may be manageable, if temperatures rise 4–5° . . . the vineyard map will never be the same again," says Bernard Seguin, head of the Climate Change and Greenhouse Effect Unit at the National Institute for Agricultural Research in Avignon. In the past thirty years, harvest dates have moved up an average of sixteen days because of unusually warm growing seasons. Grapes are reaching their sugar ripeness before their aromas fully develop, alcohol levels are soaring, and acid levels are dropping—forcing some winemakers to resort to chemistry in their cellars to produce a quaffable cuvée. Lee Hannah, a University of California–Santa Barbara based climate specialist with Conservation International, and one of the authors of "Climate Change, Wine and Conservation", adds that "as grape-growing suitability moves northward, I believe the winemaking regulations of Europe could break down quickly."

So we will see the world's wine regions remap as winemakers face contrasting experiences. At one end, we have Gerard Gauby, a winemaker in Roussillon who isn't sure that winemaking in France's southernmost regions will even survive without large-scale irrigation. If the world keeps getting hotter and drier, he says, "It's perhaps better to stop making wine, and keep the water for drinking." And at the other end, we have the exciting prospect of undiscovered terrains, as the renowned viticulturist Dr. Richard Smart made clear while addressing the Third International Sparkling Wine Symposium at Denbies Wine Estate (England). He reminds us that with rises up to several degrees Celsius on the European continent, "future sparkling wines of the quality of today's Champagne may hail from Denmark and, certainly, England." And he believes that the best vintages in Tasmania have not yet been planted—suggesting Hobart has this potential.

They tell us that the future warming trends will continue, but not for how long, nor if there will be significant cooling trends within the larger warming trend during which wine producers might perhaps relax and be able to catch their breath. How do the winemakers on the front lines see their futures?

THE VOICES FROM THE VINEYARDS: A STREAM OF CONSCIOUSNESS

Travel is fatal to prejudice, bigotry, and narrow-mindedness . . . broad, wholesome, charitable views of men and things cannot be acquired by vegetating in one little corner of the Earth all of one's lifetime.

—Mark Twain

San Patrignano is a rehabilitation community deep in the fertile heart of Emilia-Romagna. Recovering drug addicts are offered a home, food, and agricultural and viticultural training in exchange for their labor and participation in this large "family." Sitting in the enormous dining hall, with hundreds of rows of tables and benches, is slightly unsettling, even a bit overwhelming, quite Orwellian, but, it cannot be denied that this brave social model is having its successes. And the wines . . . truly well made and delicious. When questioned about the effects of climate change, their enologist made reference to NASA, which he said is his source of data.

"The most huge effect here in Italy is that the yields are getting smaller and smaller. One day, we may be able to grow Sangiovese in Alto Adige."

His neighbor, the owner at Bonzara, echoes what I have heard all over Italy. He has had to resort to machine-harvesting, ruefully adding that "it is better than leaving the grapes in the sun. Manual harvesting is best but slow and with the heat, the grapes are maturing faster and earlier and all of the varieties ripen at same time now, so time is more of the essence than ever." They are known—well, self-proclaimed—to be the first leaders in their region to lower yields by using new vineyard techniques. So I asked if the heat is now lowering their yields for them. Do they have to stop the thinning and pruning, instead, to protect the grapes? He replied that he had not really thought about it, which turned out to be not true because as we continued talking and asking questions, it became apparent that he was indeed having to factor in the heat. His assistant piped in that in the past six harvests, they officially have had their highest temperatures and their earliest harvests. He conceded this point and added that the Pignoletto was even coming in before the Merlot, which was unheard of. Then his winemaker arrived and he immediately jumped onto the topic of climate change—unprompted.

He told us that his biggest problem, no surprise, is obtaining a true phenolic maturation. He added that their vineyards sites have good water, good clones, and a good microclimate. The clones he has selected have larger leaves in warmer temperatures and so give better protection from the heat. That was the extent of his mitigation and adaptation strategy. And their Rosso del Borgo is of Cabernet Sauvignon. Cabernet Sauvignon in Emilia-Romagna. And it is 14.5 percent alcohol.

One very hot April morning in Torgiano, Umbria, I was present at a conference led by the current president of Umbrian sommeliers, Sandro Camille. He took us through the history of the earliest recorded Etruscan winemaking practices, illustrating how the vines were actually grown like trees and trellised vines, like today. He explained that Umbria comprises 8,456 square kilometers of which 75 percent is hillsides and 29 percent is in mountain areas; there are 12,189 hectares under vine. With a mild, dry climate and with hot summers, and the entire region facing south . . . he explained that the region used to grow and produce a majority of white wines, but now the opposite is true, with whites representing 43 percent and reds, 57 percent. After we tasted a series of wines, several of them being rather flabby whites and the rest, rather alcoholic reds, we started asking questions. He was reticent at first, but then conceded that they were no longer able to produce white wines with the acidity and bouquet that they once had, and they have launched "Progetto Norcia," a study to determine if they could grow some of the white varieties from Alto Adige in their highest elevations. "Umbria is a warm region, and it is getting warmer, and yes, climate change is a factor."

In France, the wine producers are revolting. They are alarmed. *Le Monde* recently warned that "French wines, elegant and refined, the jewels of our common national heritage, are in danger." Antoine Pétrus, France's" best young sommelier," says that the change in the taste of some French wines is already recognizable. "Sun-baked, unbalanced wines like those produced following the European heat wave of 2003—when temperatures soared above 40°C—were once considered a rarity. Now, they are becoming much more the norm. We've observed over the last several vintages that temperatures and effects of climate change have become far stronger." The hurdle of irrigation has been dealt with; now replanting has to have its turn. As Pancho Campo, the Spanish climate change and wine expert, reports: "I was just in Burgundy and producers there were very concerned, because they know that Chardonnay and Pinot Noir are cool-weather varieties, and climate change is bringing totally the contrary. Some of the producers are even considering starting to study Syrah and other varieties. At the moment, they are not allowed to plant other grapes, but these are questions people are asking."

And the opposite is happening in the cooler regions, like Champagne and Alsace, where ripening used to be the problem and the addition of sugar (chaptalization) was practiced in order to raise sugars. Now this practice is being replaced by acidification, the adding of tartaric acid, in order to maintain acidity levels that make these wines what they are. Olivier Humbrecht of the legendary Domaine Zind-Humbrecht laments that his father's challenge "was to ripen his grapes enough to avoid having to add sugar to the wine. Today, my problem is being able to keep enough acidity." Using tartaric acid is a practice authorized this year in Alsace by INAO for only the second time since 2003, but one that is likely to become regular very soon.

In much of Europe, grape growing is on small farms. The wine growers are farmers who are at the mercy of climate and market forces. "They will always be likely to be able to grow grapes," says Busalacchi. "The question is of what quality and what varietal." In southwest France, overproduction is a bigger issue. Busalacchi says that in regions like Minervois and Corbières, the government is encouraging quality over quantity, "grubbing up" vines by pulling them up by the roots and replacing them with other agricultural crops, and limiting new acreage. Miguel Torres agrees:

> Climate change will affect Europe more than other places. The map of the European appellations will have a dramatic change. If there is a change of 2 or 3 more degrees, where do you place a Burgundy, or a Rioja? What are you going to do? In places like Chile or Argentina, in the Southern Hemisphere, you can still go south. Here in Europe, we are demarcated by appellation of origin. If you are from Rioja, the grapes have to come from Rioja.

And if you are from the Douro, those grapes in your port have to be from Douro . . . or, should I say, those raisins? Is grape growing on the stunning slopes of the Douro going to adapt, or become impossible? Already, high temperatures have hit hard, and so far such is the style of the wine, port, that it can handle the changes in the grapes. But dry whites and dry red table wines are suffering, and so soon may Port. The intense exposure to sunlight and the prolonged droughts have stressed the vines beyond endurance, and the forecast continued heat waves and/or heavy rainfall will mean premium wines will not be possible. Like the rest of southern Europe, winemakers here are already into site, variety, and rootstock selections and trying to conserve water.

Luiz Alberto writes:

A possible solution is abandoning the south facing vineyards that are too hot (or try some dramatic canopy management changes including shading) and replant on the cooler, north facing slopes. The Valley offers 360° of exposition, but early adaptation to the new scenario is going to be key to a successful transition (and such a transition requiring new plantings will take years). Most of the new vineyards in the Douro Superior (where the rainfall level is 1/3 of the Baixo-Corgo) are already north facing. For example, the famous vineyards Quinta de Vargellas and Vesuvio are both north facing. In the wine industry things move really slowly (it takes a few years for a vine to start producing wine and many more before it starts making good wine), people have to start acting now. The rest of the world will be also responding to climate change. The efficiency of the adaptation is crucial. A region as traditional as the Douro needs to adapt quick and show flexibility. Some laws will become old and inappropriate. These laws will make no sense under the new environmental conditions and need to be eliminated. For instance, there are significant physiological and morphological differences among *Vitis vinifera* varieties and the ones that are allowed (or recommended) to be planted need to be re-evaluated in over time. There are hundreds of grape varieties in Portugal alone. The ones that are less sensitive to hydric stress and high temperatures need to be favored against the ones that don't perform as well under these conditions (such as Tinta Barroca or Tinta Francisca). However, to mitigate this issue, it's also possible to use rootstocks that are drought resistant (relatively speaking) and, because of that, R110 is becoming more and more popular in the Douro. It was already used in the past (along with 1103P), but lately more and more producers want drought tolerant rootstocks, rather quantity or quality focused rootstocks (*My Wine Studies* blog).

And although the irrigation laws in Portugal have not yet been loosened, many producers have been ignoring them anyway and either watering by hand, or using drip irrigation. But this begs my question: Why are they even planting vineyards in Douro Superior when they knew what troubles lie ahead? Alberto does not see this as a long-term option anyway: "While irrigation seems to be a solution to mitigate the problem with more frequent and severe droughts, there's a serious need to work on measures to promote the sustainability of the water supply for the entire region."

ADVID, the Association for the Development of Viticulture in the Douro, has begun a program, called ClimeVineSafe, to help winemakers with short-term adaptation strategies, one of which is to grow grapes at higher elevations (as if it not already vertiginous enough there—I've suffered too many near-death bus trips in those hills as it is). They suggest that some varieties that now thrive at six hundred meters might be planted at seven hundred or more.

Alto Adige: Room for Improvement

Crisp, personal, solid Pinot Blancs . . . Chardonnays with the muscle and salty earthiness of Meursault . . . aromatic Gewürztraminers that toy between the sharp and the sensual . . . and Sauvignons that are explosively fruity and complex . . . you would forgive me for thinking that I was in France. But no, I am in Alto Adige. And there is another surprise to come: the Pinot Nero (Pinot Noir). They are divine, and devoid of that medicinal, metallic retro-olfactive with which so many basic red Burgundies can be marked. These are fresh, elegant, and ooze a velvety smoothness of plums and warm earth, without losing the Pinot typicity and becoming jammy and sweet. And with warming temperatures, there is more to come. This is a region that is already improving with climate change.

Alto Adige, or Südtirol, is one of Italy's smallest regions (providing only 0.7 percent of Italy's total production) and can boast the fact that 98 percent of its wines are of the DOC quality category. There is archaeological evidence of viticulture here that predates the Romans, and today there are 12,500 acres of vineyards. Almost 75 percent of these are owned by cooperatives, in which, typically, each of the hundreds of members might cultivate a plot of less than two and a half acres. Cooperatives often have a negative connotation in the wine world, but not here. Here the concept works as it is meant to and produces high-quality, *terroir*-driven wines.

Nestled in the slopes of the snow-covered Southern Alps, Alto Adige has been home to the noble Bordeaux and Burgundy grapes for over a hundred years. So not "indigenous," but "traditional"! The diverse soils and altitudes welcomed them a place alongside their already established Sylvaner, Müller-Thurgau, Veltliner, Riesling, and their famously gorgeous native red grape, Lagrein. Protected by the Dolomites, the vineyards' altitudes range from 750 to 3,250 feet above sea level, and the rich soils are a geographic rainbow of dolomitic rock, fluvial deposits, porphyry, moraine debris, volcanic deposits, and slate-primitive rock.

Couple this unique climate and exposition with the quality wine-making techniques these producers embrace, and we are presented with consistent and powerfully elegant, grown-up wines that rival the French greats at half the price. With Burgundy getting a little too hot under the collar for my tastes—and they are still charging exorbitant prices—this is where Pinot Nero seems to have found the opportunity to express the best facets of its unique and elegant personality. As one of the charming producers quipped to me as I swooned over his Sauvignon: "Who needs France?"

THE ONES TO WORRY ABOUT

To cease and desist. Not a pleasant possibility. But it is very real and already happening. Drought and oversalination caused by excessive irrigation have already begun putting some of Australia's vineyards out of business. And those that are still standing are changing. Ross Brown, the CEO of Brown Brothers, has already started relocating production of his cooler varieties to Tasmania's Tamar Ridge winery. "Basically we are in the coolest part of Victoria (for wine) and that won't be cool enough to produce some of our main wines for sparkling and pinot noir." Brown says warming also presents a major challenge for wines that are already suited to warm climates, like Shiraz and Cabernet, because they will lose quality. "In a warmer climate that heat and earlier ripening period create richer and fuller-bodied wines," he continues. "But we are seeing a consumer demand for finer wines, more elegant wines and that does not augur well for people who are already making those rich fuller bodied wines."

For southern Australia, the projected warming to the year 2030 is between 0.5 and 1.5°C, with concomitant decreases in the amount of winter and spring rainfall . . . the more narrow the temperature range most suited to the phonological development and ripening of a variety, the less likely it is to continue to flourish in its present region.

Apart from this effect, the projected decrease in winter and spring rain and increase in evaporation rates in southern Australia pose a serious challenge for winegrowing, one that underlines the importance of water conservation measures such as implementing efficient irrigation practices, reducing unnecessary evaporation, improving soil water-holding capacity, minimizing water waste in the winery, and using reclaimed water when possible. Indeed, water shortage is probably the greatest threat to the security of the Australian wine industry. In addition to competing for water with other rural water users and rapidly expanding cities, the industry is faced with a lack of opportunities for dam construction and the prospect of more frequent and intense drought (Robert E. White, *Understanding Vineyard Soils*, 2009).

It is the same story here as we are hearing in all hot regions, but Australia is moving in fast-forward. In his article "Climate Change Threat to Australia's Top Wines," Robert Burton-Bradley laments the threat to the $5.5 billion Australian industry. It is now common knowledge that warming decreases wine quality, but the large bulk wine brands will be less affected than the many "smaller grape producers struggling to break even in the midst of an oversupplied market." He worries that the "costs of adapting may force some to leave the industry."

Leanne Webb, CSIRO climate change scientist and wine expert, has been studying ripening times across Australia and has found that grapes were maturing faster in recent warmer temperatures, affecting quality and taste. She confirms that some growers are already modifying their winemaking to cope with the effects; at least one major player is taking steps to move production farther south. "All my modelling is showing that if the climate warms up, given that a variety is able to ripen well in a region in a present climate, the actual quality that we see will be decreasing," Dr Webb says. "Wine drinkers may not understand the complexity that goes into it, but I am sure they can taste the difference." Dr. Webb coauthored the report *Observed Trends in Winegrape Maturity in Australia*. The CSIRO predicts temperature rises of between 0.5 to 3°F (0.3–1.7°C) by 2030, and while in some colder climates this can lead to more consistent vintages—as for Riesling in the Mosel region of Germany—in Australia the effect is generally negative. Dr. Webb and her coauthors found the average ripening creep had quickened to 1.7 days a year in the period 1993–2009 compared with 0.8 day a year from 1985 to 2009. Of forty-four blocks of vines across the dozen regions studied, only one—in Western Australia's Margaret River—bucked the trend toward earlier ripening. Dr. Webb believes it is the temperature changes combined with some vineyard management practices that are causing the effect, and while all regions will have to adapt it will be harder for the warmer zones. "Regions like the Riverina and . . . the Murray Valley for instance will be more affected than regions like Coonawarra," she says.

The large, well-known brands, known for their ubiquitous standardized products (sorry), such as Penfold's, Lindemans, Wolf Blass, Yellowglen, Rosemount Estate, and Seppelt, have been watching their vintages "ripening earlier by a day, to a day and a half, per year for the last 15 years." Burton-Bradley quotes Jioia Small, the "regional sustainability manner" for Treasury Wines, as working with the South Australian government growing vines in controlled warming experiments and adjusting their winemaking accordingly. While the results from these "climate change wines" were not yet available, she says the company's main issue was with any increase in extreme weather events like droughts and floods, which are harder to plan for. Chris Pfeiffer, from the Victorian Wine Industry Association, echoes this fear when he says that it is not so much the hotter temperatures as the erratic harvest events that are difficult to deal with . . . high rainfall and flooding are wreaking as much havoc as the heat. This is exactly what confuses people. I heard someone say, just the other day: "But how can there be global warming in Australia when they are being inundated with floods?"

CLIMATE VERSUS WEATHER: HOW CAN IT RAIN DURING A DROUGHT?

Southern Australia has a Mediterranean climate. Most of the rainfall falls in the winter half of the year and summers, in general, are dry. The 10 to 20 percent loss of autumn and winter rainfall has been made worse by the subsequent reductions in stream flow and water storages: Drier soils and vegetation soak up more moisture and therefore provide less surface runoff when rain does fall. The reduction in autumn and winter rainfall has been the theme of a series of very dry years. It is most felt in southwest Western Australia, where the last decent wet year was in the mid-1960s. This has systematically reduced inflows to dams and led to progressive depletion of water storages. For the southeast, rainfall over 2010 and 2011 brought relief after fifteen dry years. But it is important to realize that the recent rainfall was tropical in origin, and fell in spring and summer, while the long-term trend of reduced rainfall occurs during autumn and winter. In other words, the big wet year finally arrived, but the more systematic seasonal drying potentially remains. In fact, for the period April through September, it was drier than average across southern Australia in both 2010 and 2011. Western Australia continues to experience dry April–May periods.

Cycles within cycles . . .

WEATHER OR CLIMATE?

Karl Braganza, manager of the Climate Monitoring Section at the Australian Bureau of Meteorology, explains this phenomenon so well that I have included the direct link to his article (http://theconversation.com/a-land-of-more-extreme-droughts-and-flooding-rains-5184). I touched upon the difference between weather and climate change earlier, but this piece, "A Land of (More Extreme) Droughts and Flooding Rains?," is the best I have found on the subject of contradictory weather patterns and drought. It does much to help us understand the dichotomy that is Australia.

He explains these cycles within cycles and why even though the enhanced greenhouse effect means a much warmer planet, the wetter regional patterns within this warmer weather make it difficult for people to understand the long-term climatic change if they are witnessing such flooding. It does not make sense. And I am always reading articles by climate change "deniers" who write sneeringly and smugly: "I'm looking out my window right now and my neighbor's car just floated past my window . . . global warming, you say??" We know the big picture is a warming one—we just can't predict the erratic snapshots of the smaller events within.

Braganza is keen to point out the difference between rainfall and temperature. "It is clear from the public discourse that there is both a general lack of understanding of the relationship between global warming and rainfall, as well as confusion around how future rainfall projections may or may not relate to current and recent rainfall variability in Australia." As all the climate models explain, the greenhouse gases warm the atmosphere, which causes warmer oceans and a warmer atmosphere, which then holds more water vapor, which then releases more rain. So the warming and the increasing rain are one and the same, really. "The drier regions of the planet are likely to experience more severe drought interspersed by rainfall that is heavier—when it does fall. This result is known in the science as an intensification of the hydrological cycle. It has been a consistent feature of climate modelling over the last 30 years. It is the result that has been reported in every IPCC report published since 1990."

But as clear as this seems, there is still much debate over the fate of northern Australia. Will it be wetter or drier? Southern Australia forms a much clearer picture—and not a pretty one. Braganza takes a step back from the climate change scenario, for just a moment, to emphasize that the extreme rainfall events may not solely be attributed to climate change; the current weather is more intense than the usual weather patterns as just a change in the natural variability. But I should think that that is exactly what climate change is doing—exaggerating the already existing weather patterns, amplifying their effects, pushing them to extremes.

Apparently, "one of the distinguishing features of Australia's climate, in comparison to arid or semi-arid regions elsewhere on the globe, is the drenching that the entire continent receives on a semi-regular basis. For some parts of Australia, drought conditions can be caused by a prolonged lack of these very wet years alone, since those years are important to the normal hydrological cycle that ecosystems have adapted to." But the evidence is showing that the rainfall patterns are shifting, meaning when they occur along with how heavy they are. There has been a mix of events, such as some years with little or no rainfall alternating with some years of reduced autumn and winter rainfall. All this is leading "to prolonged and severe drought and water shortages." Braganza reports that "a 10–20% loss of autumn and winter rainfall has occurred over both the southwest and southeast corners of the Australian continent since around 1970. In the southeast, similar rainfall reductions have been apparent since the mid 1990s."

Australia is really heading for more changes and difficult choices. The world will be watching this nation, the country that is "the world's driest inhabited continent and regarded as most vulnerable to climate change." How ironic, or—more important—how disappointing and alarming it is to learn that Australia's new prime minister, Tony Abbott, "the man who once dismissed evidence of climate change as 'absolute crap,' has disbanded an environmental agency set up to provide independent information to the public and removed its head, the internationally renowned scientist and author Tim Flannery" ("Australia and the Climate Criminal," September 2013, *The Independent*).

Australia is afforded the limited opportunity to move south, toward the pole, where lies Tasmania, its southernmost state, where Pinot Noir, Chardonnay, Sauvignon Blanc, Riesling, Pinot Gris, and Cabernet Sauvignon are gown. Traditionally the sparkling wines have been the mainstay product here on this blustery, windy island of clay. Gregory Jones, whom you will have noticed by now is somewhat of a hero of mine, has written: "If you look at Tasmania, it was too cool to grow grapes 25 to 40 years ago. Today, it's clearly much more suitable." Warming temperatures have meant that dry whites and reds are having more success and producers are looking more at Cabernet Sauvignon, Merlot, and Shiraz. At the moment, most of the vineyards are located near Launceston and Hobart, and they have the warmer Coal River Valley and the Freycinet Peninsula regions that seem to be potential red wine production areas. My problem again, with these emerging regions, is their incredibly boring choice of internationally "safe" grape varieties. If starting "fresh," why not start "exciting"? As a New World wine region with no native *Vitis vinifera* species—and with the plethora of hundreds of varieties at their disposition to import—why go for the tried-and-true variety "brands"?

Do they still all want to try to be mini Bordeaux, mini Burgundies, and mini Alsaces? I would go for other plantings, personally, and try to make Tasmanian wine. If that can exist.

Another region moving south is New Zealand's South Island, where I have been tasting some Pinot Noirs from Central Otago that have stunning potential. They showed true Pinot Noir characteristics and were elegant and expressive. But there, too, I found some producers with 13.5 plus percent alcohol—and from this, the coolest region in New Zealand. When I asked about this, it was explained to me that at the moment the difference between day and night temperatures is so extreme that the colder nights actually shut down the vines, which are not able to process all the sugars that they accumulated during the day. Will the warming trend lessen this difference and create a more stable growing season? The soils in Central Otago are a bit different, too, rich in mica and metamorphic schists in silt loams. Most plantings are on hillsides that drain easily. But here, too, irrigation is already practiced due to the hot days and dry summers. Will climate change allow them to wean themselves off irrigation? Is that even an option, or a goal? Will we ever have a non-irrigated fine wine region again?

NOWHERE ELSE TO GO

Another New World region under the microscope is South Africa. Unlike Australia, South Africa has no more room to move poleward; it has no Tasmania, as Australia does. But those I have spoken to here, all winemakers, although aware and concerned, do not seem to feel the need to take any strong measures. They just seem to be dozily adopting a wait-and-see strategy. Perhaps it has not yet hit them where it counts most: in their wallets? South Africa is still in the top ten wine production countries, with an increase in production between 2008 and 2011 of nearly 30 percent. It is a huge bulk wine producer, so the product and the large niche it fills will not yet feel the repercussions of climate change. The consumers who enjoy South African wines are those who enjoy hot climate wine profiles, so the need to change wine style won't come from commercial market pressure, but rather from complications in infrastructure and production, such as increased energy costs and water shortages. This is, in fact, the case with all larger, bulk-wine-producing regions. They are already producing the sort of wines the European fine wine classic regions fear being reduced to producing . . . there is nowhere else to go, both figuratively and literally. So their huge bulk and lack of appellation laws will help withstand any great shifts in market profile. They will continue to make wine, any sort of wine, until they can no longer afford to. Any market pressure will be focused toward the costs of wine production on the environment, not on the resulting quality of the wine.

In the VinIntell Report of May 2012, "Future Scenarios for the South African Wine Industry," it is acknowledged that South Africa "falls within a vulnerable region as far as climate change is concerned due to its geographical location and its low level of coping capacity." South Africa is said to be one of the highest emitters of GHGs in the world, ranked nineteenth in 2005, with the agricultural sector identified as being the most vulnerable.

So the South African Fruit and Wine Initiative established the Confronting Climate Change Initiative, which announced that there would be a decrease in water availability, specifically in the southern and western parts of the country, and an increase in the frequency and intensity of extreme events such as flooding, within, overall, a general warming. But there does not seem to be much initiative being taken. The report found that the most "prominent biophysical impacts of climate change on the South African agricultural sector include a decrease in water availability, a shift in seasonal temperatures and climatic patterns and in increase in pests and diseases. The indirect impacts include an increase in energy and fuel costs, an increase in market pressure and retail demands, and the likelihood of carbon pricing in the future." I think that this is really quite removed from reality. We know that carbon pricing is not a solution if it just shifts energy use to another buyer. We know that the water will run out. We know that the costs of forcing the earth to grow something it no longer can will become unviable.

Adaptation efforts seem to be limited to cultivar changes in both rootstock and scion cultivar selection, and winery adaptation (improved insulation to reduce energy requirements; installation of solar panels to result in "cost and carbon savings and allowing a relatively fast pay-back rate of 5 to 10 years").

Dr. Wilmot James MP, the chairperson of the African Genome Education Institute, gave a lecture in 2011 on the crux of wine and climate change, arguing that the answer lies in grape genetics. "The native territory for the forebears of what became today's domesticated grapes (*vitis vinifera*) is the region that today makes up Iran, western Turkey, Armenia, Azerbaihan and Georgia, where the ancestor to *vitis vinifera* still grows wild." Working on grape genomes means there is the potential to build in greater immunity against disease and plant pathogens so wine grapes "can be grown in areas that might otherwise be inhospitable to the vine." Is this all he's got? Does he acknowledge that these native territories experienced substantial shifts in climate, thus forcing the *vitis vinifera* to migrate? He goes on to say that records from twelve weather stations in the Cape, between 1967 and 2000, show that very warm days have become warmer, particularly during the past decades.

In the future, temperatures are expected to rise in the south western Cape by about 1.5°C along the coast and by about 2°–3°C inland by 2050. How much will this affect viticulture? Vines are hardy and produce better fruit when made to struggle. But how much struggle can it take? The wine industry has been characterized by its geographical diversity, but this is threatened by climate change. When a warm region becomes hotter, diversity in the type and style of wine is more limited. On the other hand, the industry is situated in an area where there is still potential for further expansion into temperate and cool areas.

It is? Here, he is referring to coastal regions and higher altitude sites – but this potential is limited.

He quotes Suzanne Carter, an environmental and geographic scientist at the University of Cape Town: "Rainfall will decrease in the Western Cape, but by how much isn't known. Most vineyards use irrigation. Currently, most farmers don't use their full quota of irrigation water and therefore could possibly draw more water as temperatures rise. As demand for water increases, wine producers will pay more per unit of water, impacting on costs." Another UCT expert chimed in with "Grape people don't want rain, especially summer rain. They want dams and rivers full, but not rain." Ach. See what I mean? No, dams and full rivers that you can drain for your irrigation systems, further diluting your wines and further creating water shortages, are not the answer. With such warped thinking at the outset, where are they headed?

NORTH AMERICA: HOME TO BOTH EXTREMES

The United States, for obvious reasons, is one of the most difficult wine-producing countries to decipher . . . so vast is it that it possesses all the various scenarios that exist in Europe and the rest of the world. It will have within its borders regions that cease, adapt, improve, *and* emerge. Gregory Jones writes:

In North America, research has shown significant changes in growing season climates. This is evident especially in the western United States, where in the main grape-growing regions of California, Oregon, and Washington, growing seasons have warmed by 1.6°F during the last 50 years. This increase is driven mostly by changes in minimum temperatures, with greater heat accumulation, a decline in frost frequency that is most significant in the dormant period and spring, earlier last spring frosts, later first fall frosts, and longer frost-free periods. These changes have allowed a region like the Willamette

Valley of Oregon to go from a marginal cool climate region in the 1950s to 1970s to one that provides much more consistent climate and vintages today.

Dr. Gregory Jones and Stanford climatologist Noah Diffenbaugh coauthored a study suggesting that rising temperatures could reduce Northern California's prime vineyards by half over the next thirty years and that Napa Valley is approaching the temperature threshold for optimum cultivations of the grapes currently grown there. The Napa Valley is going to great lengths to argue that its microclimates are immune from what the rest of the world is experiencing. Area producers keep funding studies to say that all climate change models are exaggerated. They admit that temperatures have risen an average of 2 to 4°F (1–2°C), "mostly at night from January to August," but they deny that this has any affect on the taste of the wines, or that the consumer has noticed any changes. Yes it does, and yes we have. In fact, in all my research and conversations, this is the one region that seems to be in denial. They have a lot at stake—who can blame them? It's a $15 billion a year industry.

Terry Hall of the Napa Valley Vintners Association resists "alarmist" claims and points out that "95% of the members of the Napa Valley Vintners' Association run family-owned vineyards and make their livelihoods from viticulture and winemaking." Paige Donner in her article "Winemakers Rising to Climate Change" quotes Hall: "We're keenly aware of climate change. We're prepared to make in-field adaptations when it is necessary. But we're not there yet." Sorry, but you are. And based on my tasting notes of the past twenty years, you were ready quite a while ago. This is a region that considers a cool climate Pinot Noir acceptable at 14.5 percent alcohol.

Whenever I broach the topic of high alcohol with Northern California "premium" winemakers, they attempt to steer the conversation away from alcohol and acidity—away from the crucial components—and discuss "balance." As if to say that yes, the alcohol is high, the acidity is low, but don't worry, we extracted as much fruit as we could and added as much new oak as possible and so it all balances out. No, it doesn't. One producer recently told me that the "handprint of the winemaker is what Pinot Noir is all about." No, it isn't. *Au contraire*, and if any of these Californians had ever lived in Burgundy, they'd know that. But this is what you have to say when the heat has erased all your varietal character, taken away any soil influence there may have been, and left you with only "technique" with which to concoct a brew.

Winemakers in Oregon are equally skeptical (I have only spoken to a few) even though they will eventually benefit from warmer temperatures. But they are experiencing more extreme weather patterns and more rain and cooler temperatures, so they are not understanding that this is part of a larger warming cycle.

In his study (*Climate Adaptation Wedges: A Case Study of Premium Wine in the Western United States*, 2011), Diffenbaugh also suggests that it's possible to make viticulture far more adaptable, preserving some vineyards and traditional grapes, even as the temperature rises. "We know growers can use pruning practices to increase shading. The orientation of the vines on a given piece of land can help reduce excess heat. Irrigation and trellising practices can be used to cool the plants," he says.

Oregon and Washington were just getting accustomed to their new profile as cool climate regions. Seventy-five percent of Oregon's plantings are of Pinot Noir and Pinot Gris. They are all set up to take on Burgundy and bask in some of their southern neighbor's reflected glory . . . as California slides into unsustainability. But they are not safe, either: They have seen the length of the frost-free period increase from seventeen to thirty-five days, water shortages are now an issue, and the lack of cold means that pests and disease never die off during the winter and continue to thrive (Gregory Jones, "Climate Change and Wine: The Canary in the Coal Mine").

For a region like Oregon's Willamette Valley, or any region, the projected latitudinal shift will have an effect on varietal suitability changing the wine style. Jones uses Pinot Noir as an example: "Willamette Valley's climate over the last 30 years has, on average, been between 58–61°F, centered within the ideal climate for Pinot Noir quality and production. However, conservative projected warming rates of 1–3°F could push the region's growing season climate outside what we know today as being suitable to high-quality Pinot Noir." Oregon should be looking at some of the reds being grown in the UK, like Rondo, or warmer climate red vinifera varieties such as Tempranillo or Doletto, or the Veneto reds. It was pinned as being the US answer to Burgundy, but this is not proving true, and before they are even given another chance to convince us, they will have to look to other varieties. Washington State mirrors the Bordeaux and Rhône climates. It is trying to be both, and produces Cabernet Sauvignon and Merlot and Syrah. This is just not viticulturally feasible. You cannot do both grape climate categories justice. But before they become too settled into either niche, they already need to be experimenting with warmer reds.

And as we move even farther north, we find more fledging regions starting out and on their way. Tara Holland, an environmental scientist at the University of Guelph in Ontario, Canada, is studying the ability of vintners to adapt to the demands of a changing climate, focusing on the emerging, ten-year-old wine industry in Prince Edward County near the northeastern end of Lake Ontario. There, "freezing temperatures are currently a problem:

vintners must bury vines in the winter. If frost threatens the growing season, they now have to light fires and set up fans to blow smoke through the vineyards to create a warming thermal inversion." Jones concludes that

> for those that are able to consider new locations, areas that are higher in latitude, higher in elevation, and/or closer to the coast could potentially provide climates that would allow maintenance of style and quality. For those that are not able to consider new locations, the relatively narrow climate suitability of each variety will likely cause even small changes in climate to require shifts to different varieties or newer, more heat tolerant clones of the same variety. Fortunately, vitis vinifera has a wide genetic diversity that can enable such shifts. However, within vitis vinifera, there are few widely planted varieties that can produce quality wine in excessively warm climates.
>
> The rate of climate change and/or the rate at which variations in environmental tolerance can be exploited may therefore impose adaptability limits, particularly in long-lived systems such as premium wine, for which the long time to maturity (1–2 decades) and in-place lifetime (3–5 decades or more) increase the investment and opportunity costs of changes in location or variety, as well as the potential loss should the actual climate change be different than anticipated (Climate Adaptation Wedges: A Case Study of Premium Wine in the Western US, 2011).

It is exciting to think what new vineyard sites the future may bring. Jones agrees: "Where there is cash involved, people will make it happen. Whether they make it happen historically as we've known it, or relative to some new system in the future, remains to be seen." If the temperature increase is catastrophically extreme, though, all bets are off. "If by 2080 it's too hot to make wine in southern England, what's the rest of Europe and the world going to look like?" asks Richard Selley. "You'll have to switch to making palm wine."

Now, over on the East Coast, where Vitis vinifera has always had a bad track record, climate change means winemakers in Vermont can now ripen Pinot Noir, although their ice wine production is compromised. In North Carolina they are now growing Chardonnay, Viognier, Muscat Ottonel, Cabernet Sauvignon, Cabernet Franc, Merlot, and are looking at Syrah, Petit Verdot, Sangiovese, Tannat, and Mourvèdre as well as native and hybrid varieties such as Chambourcin, Norton, Seyvel, Vidal, Niagara, Traminette, and Chardonel because they tolerate much colder temperatures and have a later budburst, saving them from frost risk.

That sounds like a recipe for a European fruit salad, but they need to plant such a wide selection to best find what suits them. They have already weeded out Pinot Noir, Gewürztraminer, Riesling, Sauvignon Blanc, Pinot Gris, and Zinfandel. The Finger Lakes are worth watching, especially their Rieslings. And on Long Island, Chardonnay, Merlot, and Cabernet Franc are making some good inroads. Although I would like to see them play with more interesting varieties. And the wines I have tasted from there thus far have lacked acidity. As have those from Virginia.

Nova Scotia producers are looking at some of these same varieties, as they have some of the same cool climate issues. Their commercial wine industry is less than thirty years old, and the best sites in the province typically have a winter minimum above -9°F (-23°C) and growing seasons above a thousand degree days. This is truly fringe viticulture. I can remember the wind in Nova Scotia and as a child, standing on huge boulders by the sea during a storm, and nearly being blown away. Here hybrids are the mainstay, such as Baco Noir, De Chaunac, Léon Millot, Lucie Kuhlmann, Maréchal Foch for reds, and L'Acadie Blanc, New York Muscat, Seyval, and Vidal for whites. The only vinifera varieties they are attempting are Chardonnay, Pinot Noir, and Riesling.

A RENAISSANCE FOR ENGLAND

There is not the hundredth part of the wine consumed in this Kingdom that there ought to be. Our foggy climate wants help.
—Jane Austen, *Northanger Abbey*

The British Isles do not, as far as we know, possess any indigenous wine grapes (*Vitis vinifera*). There is some evidence that *V. vinifera* may have flourished during the much warmer Hoxnian Stage, which ended 374,000 years ago. There may very well have been wild grapes of other species, and there may have even existed a domestic industry. But we do know that the British wine industry became serious following the Romans' arrival in the first century AD and that the Romans brought *V. vinifera* cuttings with them, as well as Italian wines, when they invaded British shores. Grape growing and winemaking became the norm under their dominion, but we do not know if it was enough to satisfy local demand (for the Roman soldiers). British winemaking received its first blow in AD 92 when the Roman emperor Domitian issued an edict banning new vineyards in Rome as well as calling for the uprooting of most of the vineyards in its provinces.

After the eruption of Vesuvius in AD 79, which destroyed vineyards, causing a wine shortage, the Romans started replanting all around Rome, near Tuscany. Their efforts to replenish their wine trade were so successful that they created a surplus that brought down the price of their wines, but conversely, because they had gotten rid of grain fields to plant new vineyards, there was now a shortage of food. So the edict was meant to both elevate the price of their domestic wine by only having enough for domestic use and not trade, and to redress the balance of grain and grape agriculture.

I will not attempt to canvass the entire history of the English wine industry: Others have done it before me and far better than I could. Suffice it to say that I would think that without conclusive evidence of the existence of *Vitis vinifera*, one has to classify England as a New World wine country, but I think that we can safely say that it was one of the first—so it is the oldest of the new. After flourishing during both the Roman and the Medieval Warm Periods and after having been further hit by the dissolution of the monasteries and more wars, English wines are finally entering a sort of modern renaissance. We know that wine grapes produce the highest quality in the climates that are considered "marginal": As I explained earlier, this forces them to struggle for their survival as opposed to vines that are bathed in constant sunshine, irrigated liberally, and gluttonously fertilized. Again, lazy vines make boring wines. So looking to such regions in Europe for grapes is a good idea, hence the reason England grows so many "marginal" cool climate grapes of Germanic extraction.

The warming weather means that the grapes are ripening better, and unless there is a shift in the Gulf Stream, this trend will continue. English weather resembles that of the Champagne region, and English soil is composed of the same Kimmeridgian clay (calcareous clay containing Kimmeridgian limestone) found on the Kimmeridgian Ridge that threads its way through Burgundy, Champagne, the Loire, and up to the UK's south coast. This warming may very well give England the advantage, especially if the rest of Europe also keeps warming up. For example, if Burgundy keeps getting warmer, its Pinot Noir will continue to increasingly resemble the New World models. The English wine industry is undergoing some very exciting changes, and the future looks even more exciting. The soil and climate similarities southern England shares with the Champagne region of France mean that the champagne houses are buying in Sussex and Kent as potential new vineyard sites, as they deal with their continued yield losses and in their desire to continue to produce the world's leading wines. Over the past four years, England has experienced a boom in the number of hectares producing the main grapes that are grown in the Champagne region of France—Chardonnay, Pinot Noir, and Pinot Meunier. These three varieties now account for more than 50 percent of England's total varietal plantings.

"The idea that we could grow Chardonnay and Pinot Noir and get it to ripen to the level we now can was once unimaginable," says Stephen Skelton, author of the *UK Vineyards Guide*. "It's changed in 10 years, really, and that is entirely down to global warming." And it gets even more exciting when this makes England one of the hottest investment properties going at the moment. Didier Pierson, owner of Champagne Pierson-Whitaker, the first Champagne maker to purchase a vineyard in southern England, bottled his inaugural batch of English sparkling wine. And Christian Seely, director of AXA-Millésimes, who started his career in the Duoro at Quinta Noval, also launched a joint venture with a Hampshire vineyard this year to begin making a sparkling wine.

One gray cloud looms overhead, however. The UK's weather is particularly dependent upon the future cycles of the Gulf Stream. It is the only thing holding the UK in its current "false" weather pattern. If this stays where it is and it continues to warm the UK's waters and temperatures continue their steady incline, then the future of winemaking in England should continue to improve—barring more erratic wet episodes. And vineyards should find their way to the south-facing slopes of northern England and even Scotland. Henry Samuel quotes Julia Trustram-Eve: "There are as far as we know no vines yet in Scotland, although there have been rumours. It's gradually creeping up. It depends how accurate the predictions are for the long term, but some say by 2080 it will be too hot to grow grapes in southern England." ("Best Wines Will Come from Scotland if Climate Change Is Not Stopped, French Chefs Say," *The Telegraph*, Paris, August 17, 2009). But Richard C. Selley claims in *The Winelands of Britain: Past, Present and Prospective* that "if the climate in Britain cools, either through the advent of a new ice age, or, as some computer models predict, because global warming will melt the Arctic ice cap, whose waters will shut off the Gulf Stream, then viticulture is doomed." In England, at least, it seems as though it is a case of watching the weather "trends" race one another and wait to see which one comes in first place. But news just in: Christopher Trotter of Fife is about to release his first vintage of Scottish wine this year.

Another northern European region to improve with warming is Austria. Austria also has vineyards that were abandoned since the Medieval Warm Period and has already shifted into mitigation and adaptation realities. Although experiencing significant temperature increases in the past decade, this is such a cool climate region that producers will have less need for adaptation than they do for mitigation. As far as adaptation measures go, they are confident that they will not have any increasing need for irrigation, at least not in the foreseeable future, and they are mindful that this may change; the consideration of heat-tolerant and slow-maturing strains is the first step in adaptation for them.

In a recent study titled "Wine Production Under Climate Change Conditions: Mitigation and Adaptation Options from the Vineyard to the Sales Booth," the authors set out to analyze the Traisen Valley in Eastern Austria and identify which measures might be implemented in the short term that will raise the trade's awareness of the importance of sustainability in the vineyards and wineries. The data, collected from 1971 to 2008, show that a decrease in tillage intensity seems to be the most important mitigation measure during grape production, as it conserves soil carbon as well as changing packaging material, which it was determined accounted for 39 percent of the total greenhouse gas emission in the total wine production process in the region.

THE NORTHERN LIGHTS

"The Arctic is on the move. The North Pole is in the same place, but the Arctic conditions have begun to shift." According to a study in *Natural Climate Change*, conditions have shifted the equivalent of 4 to 5° of latitude southward; at the same time, vegetation has moved north, colonizing the thawing permafrost. The climate conditions of the northern latitudes now increasingly resemble those found several hundred miles farther south thirty years ago. But what sort of *terroir* lies under the permafrost? What are we going to find?

> Climate change has already brought opportunity for wine growers in Northern Europe, including the UK and Denmark. By the end of the last century, there were less than 10 commercial vintners producing wine in Denmark. There was widespread acceptance of the view that commercial production of wine here was impossible. Despite its northerly location, Denmark has been developing a wine industry over the last decades that has benefited from global warming. The Danish Vintner Association now has 1,400 members (but some are individuals with 100 vines or less), and almost 50 commercial producers. Fruit wine is obviously is another story, and Scandinavian countries have long traditions here already. Eiswein ("ice wine") might be another winner up north as it already is in Vermont (Trond Arne Undheim, *Wine, Climate and Change*).

Not surprisingly, the Danes rank about twenty-seventh in the world's league of wine consumption, and now they have joined the ranks of wine producers, too. Denmark is an emerging wine-producing country that has interesting potential. Thus far, it's taking up the habits and morays of the Old World countries, modeling its industry after the classic regions.

Increasing temperatures, with better weather in August and September and milder winters, this past decade have seen a small explosion in the number of private wineries being established with plantings of the cool climate varieties favored in England, Northern Germany, Austria, and France. In 2000, the Danish government successfully lobbied the EU wine regulation, as did Sweden and Ireland, to be included as legal wine-producing countries.

Thus Danish commercial wine growers gained acceptance to produce *vin de table*. However, EU subsidies normally afforded the traditional wine-producing countries were not granted. I wonder why? Certainly, a fledging industry capable of propping up the rest of the industry one day merits assistance?

The Association of Danish Wine Producers now has more than fourteen hundred members, and since 2007 they can carry the label "regional wine" from either Jutland (the mainland), Funen (an island), Zealand (another island region), or Bornholm (a small island in the Baltic Sea). There are no official production figures yet, as producers are trying to create a legal framework that takes into consideration factors of *terroir*, varieties, field and row orientation, vineyard elevations and steepness, soil and harvest conditions, and more. This is an interesting process, for as others are dealing with climatic changes that are making things more difficult for them, here the climatic changes are bringing new tastes, new regions, new possibilities, and better results. They are trying to keep up with the exponential growth warming temperatures bring, as opposed to decline.

So far, the climate in Denmark is allowing for a sufficiently long growing season with enough warmth leading up to harvest. Normally budburst occurs in the first part of June, and harvest is in the middle of October. So far, so classic. Average monthly number of sunshine hours (degree days) is increasing each year, and while they cannot ripen Spätburgunder (Blauburgunder, or Pinot Noir), the Danish Agricultural Ministry has permitted Akolon, Bianca, Blå Donau, Castel, Don Muscat, Dunkelfelder, Ehrenbreitsteiner, Eszter, Goldriesling, Huxelrebe, Kerner, Kernling, Léon Millot, Madeleine Angevine, Madeleine Sylvaner, Merzling, Nero, Optima, Orion, Ortega, Phoenix, Précose de Malingre Reflex, Reform, Regent, Regner, Rondo, Siegerrebe, Sirius, Solaris, Tidlig Blå Burgunder, Zalas Perle, Bacchus, Chardonnay, Maréchal Foch, Pinot Auxerrois, Pinot Blanc, Pinot Meunier, Pinot Noir, and a few others. This list is almost identical to that of England. Thus far there are no really designated or named wine regions in Denmark; they are scattered all over as they experiment and discover their own microclimates. As can be expected, their issue with sugar levels is the opposite of what the rest of the world is experiencing: They are allowed to chaptalize. The excitement is palpable in Denmark. They are enjoying their new image as a wine-producing country and know that the future looks bright . . . and warm.

I have been enjoying wines from Slovenia and Croatia. There is a great boat service from Venice, where I base myself to write, to Istria (not during the winter seasons), which means you can do a wine-tasting trip between the Western and the Eastern worlds in a day. A magical journey. The indigenous varieties such as Pošip, Babic, Teran, and the Zinfandel-related Plavac Mali make for such crisp and individual taste sensations. The whites from the Istrian Peninsula are my favorites: the Graševina and the Malvasia. And let's not forget that Slovenia's western border is Italy and that this border has had ambiguous delineations throughout viticultural history; even today, there are Italian producers in Friuli who own land in Slovenia. Southern Austria, northern Italy, and western Slovenian are soil and climate bedfellows. And the region has a future of fine wine. Although it will be difficult to know how long such a period would last before this area, too, would slide down the grape variety climate scale and into port production.

A COLD WAR NO LONGER

The Europe and Central Asia region (ECA), which includes all the former Eastern Bloc countries, Russia, and Turkey, is also facing warmer temperatures, and the risks of climate change are already being evidenced.

Average temperatures across the ECA have already increased by 1°F (0.5°C) in the south to 3°F (1.6°C) in the north (Siberia!), and overall increases of 3 to 5°F (1.6 to 2.6°C) are expected by the middle of the next century. This is affecting hydrology, with a rapid melting of the region's glaciers and a decrease in winter snows. Many countries are suffering from winter floods and summer droughts, with both southeastern Europe and Central Asia at risk for severe water shortages. Summer heat waves are expected to claim more lives than will be saved by warmer winters. And the countries most likely to experience the greatest increase in climate extremes by the end of the twenty-first century are Russia, Albania, and Turkey. So, really, the *Vitis vinifera* can never go back home, to its roots. Most countries will face a mix of losses and gains, like everywhere else, and like everywhere else, improved conditions for agriculture will be sought by moving to higher latitudes: in this case the Baltics, parts of Kazakhstan the Ukraine, and most of Russia (except for the North Caucasus). It is unlikely that at this point, they will be leaving any room for grape growing, when food production is their uppermost priority. It is a shame, as many of these countries are finding their feet again after Soviet rule. Privatization made great progress, and great wines. Romania, Bulgaria, and Georgia are all making great wine again— don't forget, this is the cradle of wine. It would be a shame if their renaissance had its rug pulled out from under it by their warming temperatures. Perhaps, along with Turkey and Lebanon, they should posture themselves as our future southern Italy or Douro Valley?

During one visit to Hungary, I was shown around a vineyard that had been returned (well, bought back at a great price) to the previous pre-Soviet owners . . . a rarity. They had not been able, however, to get their villa back, just the lands. Their vineyards had been scrubbed and planted with potatoes, cabbages, and bulk-wine grape varieties. Now, with the help of private investment, they were making beautiful wines. The first series of the tasting featured indigenous varieties, grapes that, at that point in my career, I had never heard of. Stunning. Then the owner brought out his Chardonnay. I was a bit disappointed. Why a Western, international variety? But I tasted it. It was a Chardonnay, yes, but it was a *Hungarian* Chardonnay: Hungarian first and foremost. It was original and personable and so far away from those hot, heavy, and oaky New World *examplaires*. It was divine. And I said so: Me, the one married to Burgundy for life, was seduced. I said to the owner, "This is so perfectly Hungarian—what a distinctive interpretation. Wonderful." No response from him. Then he said, "Please try my new Chardonnay. It is for the export market and I have oaked it." Hmmm. I trusted him. I tried it. It tasted as though it could have come from anywhere, even Napa. I told him this. He beamed. He giggled. He rocked with mirth. For this was his goal! I asked him why he bothered to make a wine like "this" when he is capable of making a wine like "*this*" (pointing to the first Chardonnay). And do you know what his answer was? He said, "Because I have to prove to the Americans that I can make an oaky Chardonnay as good as anyone else, before they will even look at my indigenous varieties, my real wines." Thank you, again, Robert Parker.

Brazil's southern region of Rio Grande do Sul is a relatively cool spot making sparkling wines that are gaining some attention. Ninety percent of the country's wines are produced here as opposed to the more tropical northeastern São Francisco Valley, where entry-level bulk brands are being produced. I've tasted quite a few of the sparkling wines and Merlots from Rio Grande and I would like to invoke my right to the Fifth Amendment if I may.

THE GREAT WALL OF CHINESE WINE

There is a lot of conflicting information coming out of China, and I have very little personal experience with the wines or the country—so I am looking forward to learning more. From what I have learned from my fellow Chinese writers, China hopes to become a leading world producer of "volume" wine. And London's Berry Bros. and Rudd's experts believe that China "also has all the essential ingredients to make fine wine to rival the best of Bordeaux." Speaking for the Berry Bros. website, Jasper Morris MW commented: "I absolutely think China will be a fine wine player rivalling the best wines from France. It is entirely conceivable that, in such a vast country, there will be pockets of land with a *terroir* and micro-climate well suited to the production of top quality wines." I am unable to respond to this authoritatively. I have spoken to fellow European colleagues who have visited Chinese vineyards and have reported that most (if not all?) are state-owned and state-controlled, yet there is actually very little quality regulation. As one colleague recounted: "I put my hand out to pick a grape and taste it, and the vineyard manager knocked it out of my hand and said "No! too dangerous." Apparently he was referring to the amount of pesticides and chemicals used on the grapes. So is this representative of the stuff that is going to flood our international markets?

There are more than four hundred wineries at the moment, and this number is expected to explode. A study titled *The Effect of Global Warming on Chinese Viticulture* (Hua Li, Jie You, and Xing-San Huo, 2007) tells us that, based on the analysis of nationwide meteorological data over the past forty-five years, the Chinese have mapped the variation of frost-free days (FFD), the growing season average temperature, and the growing season effective accumulated temperatures to reflect the effect of global warming. This data was applied to fourteen wine regions. It was found that in the past forty-five years, and most especially since the 1980s, the number of frost-free days increased significantly and the growing season average temperatures moved northward and westward. Their future projections showed that except for a handful of areas in the northeast, "most territory of Northern China was suitable for grapevine growing."

FUTURE *TERROIRS*: SOIL

When selecting the new vineyards in our emerging regions, the same criteria exist as have been applied since the vines were cultivated—these are, indeed, the premise of any good agricultural crop siting: knowledge of local climate, topography, and geology, the soil types, the water sources, and the site's vulnerability to pest and disease. Temperature is key to ripening and quality, yes, but there has not been enough discussion of the soils of these potential new regions. In the future, are we going to be simply so thrilled to have found a place where wine grapes grow that we will be willing to disregard all other variables in the quality equation?

Soil Profile Properties Relevant to Site Selection

Soil depth—depth to a compacted layer	Affects rooting depth and the possibility of waterlogging: may indicate a need for subsoiling
Color	Orange to brick-red (resulting from Fe oxides) indicate good drainage; dark colors indicate organic matter; "mottling" indicates periodic waterlogging
pH	Affects the availability of several nutrients; extremes of pH inhibit root growth
Texture, stones, and gravel	Influence a soil's water-holding capacity, drainage, structure, and ease of cultivation
Structure	Influences soil stability when wet, hard-setting when dry; also aeration, drainage, and ease of root penetration; cracking clays swell when wet and shrink when dry
Consistence	Describes the strength and bulk coherence of a soil, which influence drainage, root penetration, and ease of cultivation
Presences of limestone of chalk ($CaCO_3$) and its hardness	Indicates neutral to alkaline pH; usually associated with good structure, but can restrict rooting if very hard
Rippable rock or impeding subsoil layer	Influences rooting depth and soil drainability; if fractured, probably can be subsoiled

Robert E. White, *Understanding Vineyard Soils*, Oxford University Press, 2009, page 37.

Site Selection: The Checklist

Climate

- Grape growing is limited by certain climatic conditions: A reasonably long growing season (150 to 180 days) with relatively low humidity (rainfall of less than 800 millimeters per year) but sufficient soil moisture is necessary.

- The temperatures from April to September (Northern Hemisphere) are crucial for reaching good development of the vine and ripening of the fruits. When temperatures are below 50°F (10°C), vines are dormant. The optimum temperature is between 59 and 77°F (15–25°C). Temperatures higher than 82°F (28°C) will stop growth.

- Frosts occurring after vine growth has started in spring could kill off most of the fruitful shoots and reduce the harvest to nil.

Soil

- Fertility is not as important as soil structure. This is climate-dependent. A cooler region can handle more fertile soil, whereas in a hotter climate, too much organic matter will create problems controlling vegetative growth.

- Different soils are better in different climates. Hotter climates need clay loams, or clay, with less than 50 percent stones, so to maximize water retention. In cooler, wetter climates, gravel soils absorb heat and warm the soils, and sandy, light clay soils with a high proportion of gravel and stones are most desirable as they allow excess water to drain; but alkaline soils must be avoided.

- Good drainage is very important. Stagnation causes humidity, which attracts disease and pests.

Aspect

- Hillsides and slopes, not valleys. Valleys give vines too much access to sun and winemakers are too tempted to fertilize, irrigate, machine-harvest, and use wider spacing, which means the vines' roots stay shallow. Slopes push cold air downward and away, which helps protect the vines from cold and frost, they drain water better, and they force the vines to cling to the hillsides, driving their roots ever deeper to find nutrients, water . . . and *terroir*. Traditionally, south-facing slopes are the most sought-after expositions to have in the Northern Hemisphere. This is because the best part of the sunlight hits the vines at the right part of the day. Although now, with the rising temperatures, where possible, some wine growers are planting vineyards to face northward, to protect from sun exposure.

WE WILL HAVE TO CHANGE OUR IMPRESSIONS

One aspect of adaptation will be the need to remarket the changes in the wine world: the new places, the new grapes, and the new tastes. The new brands. Past images and past definitions will have to be remade and rewritten. As Gregory Jones points out, it will not be an easy task: "There may also be adaptation potential through changes in marketing, which can strongly influence the public's perception of 'quality wine.'" But he is less confident that a good marketing strategy, or anything, can convince the consumers to accept a lesser wine quality. Meaning, they already have their expectations of what a

Chardonnay, or a Pinot Noir, or a Sangiovese should taste like. I would argue that the New World models of the classic Old World, cool climate grapes have already done this for us. Many of these classic profiles have already changed toward bigger and bolder styles—both those produced in the New World as well as those produced in the warming Old World. We need only look at the New World interpretations of these grapes to see the future of the changing Old World. Consumers already seem content with these versions. They would likely welcome any attempts to find new, cooler homes for these migrating models. They ask only for "good, well-made wines," and frankly, in my opinion, this definition is already sufficiently wide.

There is too much emphasis placed on the "subjectivity" of wine quality. I hope that the sources sited in this work have thus far have convinced readers that there are objective qualitative benchmarks by which to measure a wine. Too often I hear: "Oh, wine experts. What a lot of phonies. It's about what you like. If I like a wine, then it's a good wine." Usually this is a viewpoint being espoused by an overweight, middle-aged man clad in red corduroy trousers holding court at the club in Henley. Yes. If you enjoy a particular wine, if it is to your taste, then that is a good thing. But it does not then follow that it is a well-made wine. I can enjoy a chilled glass of an overly oaked, Juicy Fruit chewing gum New World version Chardonnay while floating in the pool, listening to the Eagles. I know that it is flawed, unbalanced, but who cares? I'm enjoying it. If we only drank mature Premier Cru Meursault or Grand Cru Puligny every day, we'd lose our appreciation for it. And superimposing this thought onto the replanting discussion: Will we mind what our Sancerre is made from, as long as it is well made and we enjoy it? And as noted earlier, how many consumers know which grapes are grown for which wine appellations anyway? We want good wines. What is a great wine, anyway?

Besieged by the great names all the famous châteaux, don't you wonder what is behind their greatness? And how do you know a good wine from a great? And what makes a great wine great? Should technical composition take precedence over artistic expression? Should we revere creativity, or classicism? Do we commend effort, or merit? Do we demand qualitative consistency, or adapt to natural variations? Typicality or originality? Tradition or invention? Or do we leave it strictly to personal taste?

The answer: all of the above. Every one of these criteria overlaps onto the others until they become one and it is no longer possible to identify, dissect, or classify. A great wine is just grand. A great wine is one that responds to all the technical demands of winemaking without losing its soul. It is capable of seducing without offense. It has the breeding and tradition that already inspire deference, yet it continues its efforts to please. It is reliable but never boring. It reflects the origin of its beginnings without neglecting its individuality, and it is the quintessential marriage of ancient and modern. The definition of true greatness is universal and will remain unchanging. Great wines do not bend at the will of consumer trend; they endure change and, in their timelessness, shape their epoch.

THE OBJECTIVITY OF GREATNESS

How can we conserve greatness when the elements that create it change? When a *terroir* loses its specificity? When the rain and the sun and the wind so drastically alter their course? When mountains appear where once there were valleys and where once there were rivers there are deserts? The magic trilogy of soil, climate, and grape variety must be considered the incontestable and objective indication of a wine's quality. It has been proven time and time again that a particular grape variety when married to a particular soil in a particular climate produces a unique and specific result. There are some grapes, soils, and climates that will never be able to produce a great wine, no matter what intervention humans offer to remedy the situation. The bonds that secure this trilogy are quickly breaking down before us.

THE SUBJECTIVITY OF GREATNESS

What amount of responsibility should the wine producer be given? His or her influence is subjective, yet there are objective criteria in making a great wine. Humans subjectively set in motion certain fundamental truths in the elaboration of a wine. Every step of winemaking, from planting the vines to bottling the wine, requires a decision that will affect the quality of the outcome. It is no coincidence that those wines born from those choices that are the most difficult and the most expensive to make will be greater wines and therefore of greater value.

Then there is the question of personal taste. Not everyone will share the same inclinations. Taste is an odd animal. For although there is the fact that we share inherent and permanent preferences at birth (for sweetness, for example), taste is also variable within time. We know that our Greek ancestors preferred spicy, resiny wines, and that early white wines were drunk oxidized and were rather acidic. But instead of insisting that this was because of the collective taste at the moment, let us remember that this style of wine was preferred only by the general public, and that the average winemaker did not have the same winemaking techniques or capabilities that we have today. We know that a privileged class already existed whose members were producing and drinking aged, bottled wines that more closely resemble our wines of today. Perhaps then it wasn't collective preferences that created the product, but the fact that the collectivity learned to like what was most readily available and most easily produced. Who is to say that if a bottle of Romanée Conti 1967 could have been made and tasted, it would not have been preferred?

Is it safe to surmise that great wines are, by definition, universally appreciated because of their classic construction? Universal classicism is not to be confused with the modern viticultural globalization, where the objective is to create a wine that appeals to all. The difference is that a great wine remains unchanged and compels the common voice to rise up in appreciation, whereas a less great wine caters to the taste of the majority and seeks a commercial and gustatory common denominator. But now the great wines *are* changing and the climatic parameters that we are being given will mean that the commercial, international wines popular with the mass market will become the norm—they will be all that we are able to produce. Will that change our definition or our preference for the classics? We have already entered the "post-classic" era of wine. With the higher alcohol levels, the lower acidity, and the herbaceousness brought on by vigorous vegetation (brought on by higher levels of carbon dioxide, and greater reliance upon irrigation), our wines have already changed their classic taste profiles. Will we forget them? Will they forever be erased from our palates, and will we be forced to accept lesser wines as our "future classics"—or is that a viticultural oxymoron?

Future Facts Page

- The IPCC forecasts a temperature rise of 2.5 to 10°F (1.4–5.6°C) over the next century.

- The average global temperature will increase by a further 3.2 to 7.2°F (1.8–4.0°C) by the end of this century, as compared with 1990 (IPCC).

- France's 2003 July–August heat wave was likely the warmest summer in parts of central Europe since at least 1540.

- The 2006 European heat wave had French June–July temperatures significantly above average, with July as the second hottest month on record. The 2009 heat wave had August temperatures reaching 97°F (36°C) in the North and 104°F (40°C) in the South. And during the 2009 December cold spell, temperatures fell to -4°F (-20°C) across France (Met Office).

- In the United States, wine must have a minimum of 7 percent alcohol and a maximum of 14 percent (with a legal variance of 1 percent) to be legally considered wine. More than 15 percent alcohol by volume and the wine is taxed as a fortified wine (port is 18-20 percent).

- In Europe, the legal ranges of alcohol by volume for wine are 8.5 percent to 14 percent, with regional laws further regulating alcohol content. Variances in alcohol content may only be rounded to either the closest whole or the closest half unit.

- In Australia and New Zealand, the maximum amount of alcohol allowed is 22 percent. However, these wines could not be sold as table wine under EU or US laws.

- Global warming has been around 1.2°F (0.7°C) over the last hundred years. Warming has occurred almost twice as fast during the last fifty years compared with the hundred-year period as a whole (IPCC).

- Most climate change scenarios indicate that the highest and most rapid temperature increases will occur in the Arctic, subarctic and alpine regions (IPCC).

- Temperature will rise more in Scandinavia than the global mean. A rise in mean temperature in Sweden has been estimated of about 1°F (0.4°C) per ten-year period (IPCC).

- In North America, there will be decreasing snowpack in the western mountains, a 5 to 20 percent increase in yields of rain-fed agriculture in some regions, and an increased frequency, intensity, and duration of heat waves in cities that currently experience them (IPCC).

- In Latin America, we will see a gradual replacement of tropical forest by savanna, in eastern Amazonia; a risk of significant biodiversity loss through species extinction in many tropical areas; and significant changes in available water (IPCC).

- Europe will have an increased risk of inland flash floods, more frequent flooding, and increased erosion from these floods as well as sea level rises. There will also be more glacial retreat in mountainous areas, reduced snow cover, and heavy reductions of crop yields in the south (IPCC).

- By 2020, there will be between 75 and 250 million people in Africa directly affected by water stress and crop yields decreased by 50 percent (IPCC).

- By 2050, freshwater availability will significantly decrease in the South, Central, West, and Southeast. More flooding will devastate coastal areas and disease from both floods and droughts will increase death rates in other regions (IPCC).

- The IPCC states that it is "virtually certain" that there will be a contraction of snow cover areas around the globe along with an increased thawing of the permafrost regions and a decrease in sea ice.

- We will continue to experience an increased frequency of hot extremes, heat waves, and heavy precipitation. And heavy precipitation will increase at the higher altitudes while decreasing in subtropical land regions.

- The IPCC and its sources have "high confidence" that there will be a continuation of decreased water resources in many semi-arid areas including the western US and the Mediterranean basin.

- The Loire Valley, from 1960 to 2010, saw significant increases in mean temperature (by 2.3 to 3.2°F/1.3 to 1.8°C) over the growing season (April to September) throughout the valley, with maximum temperatures increasing more strongly than minimum temperatures.

- By 2040, there could be 50 percent less land suitable for cultivating premium wine grapes in high-value areas of Northern California. In Napa the average temperature could increase by 2°F (1.1°C) with ten more "very hot days" per growing season, thus reducing the amount of land suitable for growing Cabernet Sauvignon, Pinot Noir, and Chardonnay (Woods Institute for the Environment, Stanford University).

- In Germany's Bavarian regions, warmer seasons from 1949 to 2010 have resulted in greater ripening potential for Müller-Thurgau, Riesling, and Sylvaner grapes. The sugar content increased while the acid component remained constant, resulting in a changed grape composition that has the potential to alter wine typicity and quality.

- Europe shows a strong contrast in projected precipitation changes, with large decreases in the south and large increases in the north. France falls toward the southern region with decreasing precipitation, projected decreases of up to 20 percent in the southwest, and smaller decreases of up to 5 percent farther north (IPCC).

- Recent droughts in the Po River Basin (north Italy) in 2003, 2005, and 2006 have highlighted that the north is susceptible to severe droughts. There has been widespread warming over Italy since 1960 with greater warming in summer than winter.

- Precipitation is projected to decrease over most of South Africa, particularly in the far west, with decreases of up to 20 percent. Projected temperature increases over South Africa are up to around 7°F (4°C) inland, and around 5°F (3°C) nearer the coasts.

- A widespread warming trend has been recorded over the United States since 1960. There has been a widespread reduction in the number of cool nights across the whole of the contiguous US, with stronger decreases observed toward the coasts along with a widespread increase in the number of warm nights. There has been a general increase in summer and winter temperatures averaged over the country.

- In the United States, there has been a positive trend in precipitation on the northeastern seaboard since 1960, except for the remainder of the country, where there is no significant trend. Climate change project temperature increases over the USA to be generally higher in central, southwest and northern regions, of up to around 7 to 8°F (4–4.5°C).

- The United States shows a north–south division in projected precipitation changes, with the highest decreases, of up to around 20 percent, in the Southwest, and increases of up to 10 percent in the Northeast. Uncertainty arises among the studies due to the fact that the US is sandwiched between strong increases in precipitation over Canada, and decreases over Central America and the Caribbean (IPCC).

- England and Wales have doubled their number of wineries to over four hundred in less than a decade (*Daily Mail*, London).

- China, the world's fifth largest wine producer, is on track to double its grape production in five years to become the world's largest producer (Wine-Searcher). Chinese wine consumption will double by 2016 to four hundred million cases a year, making the country the world's largest consumer.

- Global consumption grew by 1 percent in 2012, while wine production fell by 5 percent to 2.8 billion cases due to bad weather in Europe and Australia (Morgan Stanley).

LET'S DRINK TO THE FUTURE

All the projection models I have researched gear their projection periods either to 2050 or to 2100. *Erratic* and *intensive* are the two most evident common denominators threaded throughout all the analyses. Rain is falling harder and longer; the sun is beating hotter; the wind howling stronger; and the earth is gasping to keep up, its self-regulating mechanisms being pushed to their limits. We are not helping: This is true. And wine grapes are far from being the only victim. In Vietnam, deforestation and the destruction of natural habitats have left the country even more vulnerable to the forces of nature. When nature goes "out to play" it expects to encounter other forces of nature, to meet on a level playing field, not to be confronted with an opponent compromised by an unnatural and abused landscape. The poor farmers in Vietnam are frantic to grow as many low-grade inferior coffee beans as possible—the stuff of the Nescafé instant coffee sachets we find in our hotel rooms when we fly across the world for a wine tasting. They grow this golden bean at the expense of other, more essential food crops—such is its value in the market. They can get three times the money they do for corn, and they are frenetically razing forestland, over-irrigating, and overfertilizing to capitalize on the resulting higher yields, unwittingly opening themselves to more disaster and destruction and hurling themselves toward an unsustainable future. All the while complaining that the unpredictable weather is ruining their crops! And around and around it goes . . .

Erratic and intensive applies perfectly, too, to the vineyard puzzle that is forming before us. The wines of the Southern Hemisphere will at first do better due to their coastal influence. But we will see less fine wine, fewer *terroir* wines in general unless newer regions, hopefully, reveal new *terroirs*. We don't know where the future "Chablis" will be. I think that those regions in northern Europe will be the first to find our next classic regions and *terroirs*. The Old World *terroirs* will hang on for as long as they can with their current *encépagement*, trading on their established appellation "brands" until with forced irrigation and heat they become even greater New World caricatures of their old selves than they are now. Full-scale replanting programs will eventually be embraced, first by exploring the forgotten indigenous grape varieties and then by adopting others from other regions, or creating new ones, as the climate scale moves north. Plantings at higher elevations within the appellations will initially expand planting areas, as will the development of new varieties for these regions. Many have already begun to implement some or all of these practices, both officially and unofficially. Others are adopting a wait-and-see policy.

This is my first attempt to try to make sense of the changing wines around the world. Both wine and climate change are exhaustive subjects in their own right. I could not hope to delve into all the detail I wished. I do hope that this first work will provide a modest contextual background through which to view future changes. And again, because it is a forever changing and renewing subject, I am looking forward to studying some aspects in greater detail—say, specific *terroir* analyses to assist in divining our next Pouilly-Fumé or Rioja. Which begs the question: Should we look for, will we ever find, taste profiles comparable to our great classics, or should we simply accept whatever Mother Nature sends our way? I should like to produce detailed country profiles, providing very specific regional mappings and projections. And I am also currently working on a modified version of Bordeaux's *en primeur* system. There is so much to think about and I am convinced that there is also much to look forward to—if we acknowledge and embrace the changes . . . and prepare for them. So how do we prepare for the unpredictable?

European wine regions need to break away from their regulatory bodies if these do not soon start supporting adaptation efforts. At the moment, theses bodies are misguidedly regulating the wrong efforts. They think that they are safeguarding quality, but they are stifling change and a move toward future quality. Loosening planting rights, both regarding what to plant and how much to plant, won't lead to a sort of free-for-all and an abuse of quality. There are already those wine producers who flaunt the rules and make substandard wines within the official current frameworks. Appellations need to be better used as brands possessing great strength. It won't matter if Sancerre, for example, is eventually made from Trebbiano, or Viognier or Chenin Blanc. It just has to be well made and refreshing. And a Chablis Grand Cru, for instance, with 13.5 percent alcohol is neither. There will still be good Sancerre and Chablis producers and bad Sancerre and Chablis producers—just as exist today. The Old World will have to have the confidence to reinvent itself. It helps that most international consumers do not know that Chardonnay is grown in Chablis, Nebbiolo in Barolo, and Tempranillo in Rioja. This ignorance can be turned into a truly blissful advantage. Just give them good wines, wines that are personable, refreshing, interesting, and that express the great *terroirs* in a different but equally masterful voice. For example, the great châteaux of Bordeaux need not close up shop. These wines are no longer what they used to be and they need to accept this and find a new way to communicate what nature is giving them now.

Those vineyards around the world that are at higher altitudes and near the ocean will have a head start (for a while) and will benefit from more consistent growing seasons and growing days. While the bulk wines from warm-climates will also struggle on, their quality will remain unaffected until yields are made impossible due to heat and drought or over-irrigated, salinated soils, or due to exorbitant energy and operating costs. Those in the Southern Hemisphere who can find cooler climates closer to the pole will be the first to have to deal with their extreme day–night differential, which may decrease with warming, before they can speak of fine wine possibilities. Ultimately, the Southern Hemisphere will lose the race as it runs out of space in which to move southward. Then of course, there are those who say that rising sea levels will obliterate those outlying coastal regions that will not already have been lost to drought and heat. But this is an eventuality too far away to worry about now. I hope. I think that we already have enough on our plates.

In chapter 1, we looked at how climate events shaped plant, animal, and human migration and played a hand in bringing down great civilizations. We are meant to learn from our past mistakes. We hear this often enough. And here we are being given a perfect opportunity to use past templates to chart our future. A few years ago, when a group of scientists went off to search for the lost city of Ubar in the Arabian desert, they enlisted the aid of NASA, who let them take photos from the space shuttle. These revealed a series of roads, long buried under mountains of hot sand, that all seemed to merge at a spot in southern Oman that was once evidently an oasis haven: their lost outpost. We also know that there was a great Egyptian kingdom that collapsed during a long drought over 4,200 years ago. Drought also played a part in bringing down the Mayans in AD 900 and Angkor in Cambodia in the 1400s. Dr. Jason Ur of Harvard University, an archaeologist specializing in ancient Mesopotamian cites, explains that "when we excavate the remains of past civilizations, we rarely find any evidence that they as a whole society made any attempts to adapt in the face of a drying climate, a warming atmosphere or other changes. I view this inflexibility as the real reason for collapse." A powerful thought. Does this mean that they found no evidence of irrigation, of dams, or of different vegetation patterns? Once drought takes hold, does there not exist a finite number of adaptation measures before life, all life, becomes unsustainable? Still, this message from the past should be heeded.

We should look upon this new phase as an exploratory adventure into new tastes and pleasures. We also need to support those who are most affected by these changes, those whose lives will be forever altered, for better and for worse—the wine producers—and to assist in keeping their enterprises viable as they adapt and experiment. Viticulture's satellite industries will be less affected, and differently so, for there will always be wine to bottle, to cork, to label, to ship, to buy, to promote, to sell, and to drink. As viticulture areas increase and decline both in volume and in surface area, a balance will be found. For example, what is lost in Napa may be gained in the Rockies. The vinous tide will ebb and flow, until either nature or humans stem it. Like the poor coffee growers in Vietnam, viticulturists will also face the day when they have to ask themselves, or answer to others, whether or not we can afford such luxuries. There will be wines that we will miss, and wines that we will welcome. Just as with good friends.

In a recent conversation with Professor James Lovelock at the Oxford Literary Festival's launch of his new book, *A Rough Ride to the Future*, he assured me that his previous alarmist predictions were just that, and that he now believes that the global warming is slower than he originally stated, but still, very real. A wine lover, he confirmed his belief that Bordeaux will go the way of southern Spain and suggested that wineries across the globe adapt by relocating to coastal regions, reminding me that the world is three-quarters water. "It [wine] is going to all be very different. But we just don't know what's going to happen." When I asked him if the subject of this book is already moot, if preserving the wine industry is too frivolous an endeavor when we are being told that there will be a day when we have to choose between using water to grow corn, cotton or cabernet, he insisted that it was not as urgent as it is made out to be. "Don't worry", he said, "That's a long way away. We have time". Let's drink to that.

Int.   Terrace of a Stone Villa, Norris Geyser Basin, Yellowstone Park, circa AD 2080—Night

A warm, late-summer evening. The last rays of the sun illuminate the surrounding lush garden, laden with fruit trees and flowers. Mount Rushburn, though twenty-eight miles away, looms overhead, a fertile Green Giant. A couple are preparing for a dinner party. She is setting out bowls of fresh figs, almonds, olives . . . He is carrying wine bottles to the table. He is dressed in a terry-cloth robe.

EVE

Darling, you haven't dressed yet. The guests will be arriving soon. Where have you been?

ADAM

I was in our hot-geyser room. I really needed to unwind with a steam bath.

EVE

Lucky you. I've been slaving over this meal all day.

ADAM

Well, it smells great—what's on the menu?

EVE

All your favorites. I'm starting with canapés of sashimi of trout and bison blinis with sour cream. The first course is elk carpaccio and the main will be black bear casserole.

ADAM

Wow. Great choices. We still had all that?

EVE

Yes, from the deep freeze. But don't tell anyone.

ADAM

Well, they'll guess anyway. No one's seen a live bison in Yellowstone for years. So, which wines do you want with that?

EVE

Well, I went into the market this morning and picked up some of that local Greco di Tufo from Signore Vesuvio's new vineyard. Our volcanic soils give it such lovely acidity and minerality. Perfect for cutting through that trout, don't you think? We need something for the cheese course, though.

ADAM

Absolutely. I've had a look in our vinoteca. We seem to have finished our last case of that 2020 Napa Valley port, and I don't think I'll be able to get my hands on any more. Way too expensive.

EVE

God, how I miss the days when we could import any wine from anywhere we liked. Remember that Danish Malbec we fell in love with on holiday last year?

ADAM

I know, but how about a bottle of that delicious Nebbiolo from California's Central Coast?

EVE

No. None left in the shops. And I heard that they have gone out of business. Couldn't afford their water or solar bills anymore.

ADAM

That's a shame. So what's for dessert, then?

EVE

(grabbing onto the edge of the table) Darling, did you just feel something?

THE END

BIBLIOGRAPHY

BOOKS

Bordeaux et ses Vins (Guide Féret), Marc-Henry LeMay, 1995

Bordeaux: People, Power and Politics, Stephen Brook, Mitchell Beazley, 2001

Climate Crash, John Cox, 2005

Commonsense of Wine, The, André Simon, 1966

Dégustation, La, G. Fribourg, C. Sarfati, 1989

God Species, The, Mark Lynas, 2011

Great Vintage Wine Book, The, Michael Broadbent, Mitchell Beazley, 1980

Green Illusions, Ozzie Zehner, 2012

Handbook of Fruit Science and Technology, Salunkhe and Kadam, 1995

High Tides, Mark Lynas, 2004

Histoire de la Vigne et du Vin en France, Roger Dion, 1977

Inconvenient Truth, An, Al Gore, 2006

Naturalis Historia, Pliny the Elder, circa AD 77–78

New World Wines, Julie Arkell, London, 1999

Oenologie et Crus des Vins, Piallat and Deville, 1983

Pairing Wine and Food, Linda Johnson-Bell, 2012

Parallel Lives, Plutarch, circa AD 110

Rough Guide to Climate Change, The, Robert Henson, 2011

Short History of Wine, A, Rod Phillips, 2000

Six Degrees, Mark Lynas, 2007

Terroir, James E. Wilson, Mitchell Beazley, 1998

Understanding Vineyard Soils, Robert E. White, 2009

Understanding Wine Technology, David Bird, 2000

Vintage: The Story of Wine, Hugh Johnson, 1989

Wine and Vineyards of France, Jacquelin and Poulain, 1962

Wine, Terroir and Climate Change, John Gladstone, 2011

Winelands of Britain: Past, Present and Prospective, Richard C. Selley, 2004

WEBSITES (RESEARCH CENTERS, NEWS, STUDIES, BLOGS, PRESS)

Note: I also used numerous regional body websites, but they are too numerous to list here.

Argentina Wine Guide: www.argentinawineguide.com

Australian government's climate change site: www.climatechange.gov.au

Bilan Carbone Method: www.eco2initiative.com

Bordeaux Vintage Guide 1900 to Today: www.thewinecellarinsider.com

California Sustainable Winegrowing Alliance (CSWA): www.sustainable winegrowing.org

Carbon Fund: www.carbonfund.org

Carbon Neutral Kawarthas: www.carbonneutralkawarthas.ca

CCSP: www.climatescience.gov

China Wine Institute: www.winechina.com

Climate Central: www.cliamtecentral.org

Climate Partner: www.climatepartner.com

Climate Project: www.climateproject.org

Climate Reality Project: www.theclimaterealityproject.org

Climate Reduction Challenge: www.crchallenge.org

Commonwealth Scientific and Industrial Research Organisation (CSIRO): www.csiro.au; www.climatechangeinaustralia.gov.au

Confronting Climate Change: South Africa: www.climatefruitandwine.co.za

Daily Climate: www.dailyclimate.org

Dr. Richard Smart: www.smartvit.com.au

Environment Canada: www.ccsn.ca

Finger Lakes Wine Alliance: www.fingerlakeswinealliance.com

Global Carbon Project: www.globalcarbonproject.org

Hadley Centre: www.metoffice.com

Harpers: www.harpers.co.uk

Institut National de l'Origine et de la Qualité (INAO): www.inao.gouv.fr

Intergovernmental Panel on Climate Change (IPCC): www.ipcc.ch

International Organisation of Vine and Wine (OIV): www.oiv.int

Kyoto Protocol: www.thekyotoprotocol.com

Mark Lynas: www.marklynas.org

NASA: www.climate.nasa.gov

National Oceanic and Atmospheric Administration (NOAA): www.climate.gov

North Carolina Wine: www.ncwine.org

Oxford Environmental Change Institute: www.eci.ox.ac.uk

Oxford Radcliffe Science Library: www.bodleian.ox.ac.uk/science

Pew Center on Global Climate Change: www.pewclimate.org

PNAS: www.pnas.org

Professor Gregory V. Jones: www.sou.edu

Rapid-Watch: www.noc.soton.ac.uk/rapid/rw

Real Climate: www.realclimate.org

Skeptical Science: www.skepticalscience.com

Stockholm Environment Institute: www.sei-international.org

Sundrop Farms: www.sundropfarms.com

UK Climate Impacts Programme (UKCIP): www.ukcip.org.uk

University Corporation for Atmospheric Research (UCAR): www.ucar.edu

US Global Change Research Program: www.globalchange.gov

Wine and Climate Change Portal: www.wineandclimate.com

Wineries for Climate Protection: www.wineriesforclimateprotection.com

World Bank Climate Change Knowledge Portal: www.worldbank.org

World Climate Organisation: www.worldclim.org

STUDIES AND LEGISLATION

Adaptation to Climate Change in Europe and Central Asia, World Bank, June 2009

Climate Adaptation Wedges: A Case Study of Premium Wine in the Western US, Diffenbaugh et al., 2011

Climate and Phenology in Napa Valley: A Compilation and Analysis of Historical Data, R. Cayan, K. Nicholas, M. Tyree, and M. Dettinger, 2011

Climate and Terroir (Paper 4), Gregory V. Jones, Geological Society of America's Annual Meeting, 2003

Climate Change in the Vineyards: The Taste of Global Warming, Gregory Jones, Southern Oregon University

Climate Change, Wine and Conservation, Lee Hannah et al., PNAS, 2012

Climate: Public Understanding and Policy Implications, House of Commons Oral Evidence, Science and Technology Committee, September 2013

Code of Federal Regulations, Alcohol, Tobacco and Firearms, Part 9: American Viticultural Areas

Douro Region Wine Cluster: Climate Change Adaptation, ADVID, November 2008

"Earlier Wine-Grape Ripening Driven by Climatic Warming," L. B. Webb et al., nature.com, February 2012

Effect of Global Warming on Chinese Viticulture, The, H. Li et al., College of Enology, June 2007

EU System of Planting Rights, The, Directorate General of Agriculture and Rural Development, April 2012

Future Scenarios for the South African Wine Industry, VinIntell, May 2012

Guide to Greenhouse Gas Reduction for South Australian Grapegrowers and Winemakers, November 2010

Guide to Wine Law, UK Food Standards Agency, April 2013

Heating Degree Days, Cooling Degree Days, and Precipitation in Europe, Rasmus Benested, 2008

Impact of Climate Change on the Margaret River Wine Region, The, Roy Jones et al., 2010

Intergovernmental Panel on Climate Change, 1995, 2007, and 2013 reports

International Organisation of Vine and Wine (OIV): Greenhouse Gas Accounting Protocol, 2011

Irrigation Water Requirements of Wine Grapes in the Sonoita Wine Growing Region of Arizona, Slack & Martin, University of Arizona, 1999

Methodology of the Geoviticulture MCC System, Embrapa Uva e Vinho, June 2008

"Naturalised Vitis Rootstocks in Europe and Consequences to Native Wild Grapevine," Arrigo and Arnold, PLOS ONE, June 2007

Oregon Climate Assessment Report, Oregon Climate Change Research Institute, December 2010

Perceived Benefits and Costs of Sustainability Practices in California Viticulture, The, Mark Lubell et al., UC Davis, 2008

Projected Shifts of Wine Regions in Response to Climate Change, Moriondo and Jones et al., March 2013

Role of Planting Rights for the Future of the European Wine Sector, The, Copa-Cogeca, 2010

South American Viticulture, Wine Production and Climate Change, Canziani and Scarel, Universidad Catolica Argentina, 1990

Study: Extreme Heat Reduces and Shifts US Premium Wine Production in the 21st Century, White et al., 2006

Vineyard Cover Crops and Tillage Practices, Dr. Kerri Steenworth

VinIntell report: Future Scenarios for the South African Wine Industry, May 2012

Vulnerability of the Grape and Wine Industry to Climate Change: A Study of the Okanagan Valley BC, Suzanne Belliveau, September 2010

Working Paper: Wine Production in Denmark, Bentzen and Smith, Department of Economics, September 2009

CONFERENCES

Climate Change and Wine 2011, Spain: www.climatechangeandwine.com

Ecososteniblewine 2012, Spain: www.ecososteniblewine.com

Future of Sangiovese Symposium, Petra Winery, Tuscany, 2009

IFSA Symposium 2010, Austria: www.ifsa-europe.org

Oevioviti International Symposium "Alcohol Reduction in Wine," Bordeaux, 2013: www.fondation.univ-bordeaux.fr

OIV World Congress of Vine and Wine (annual): www.oiv.int

United Nations Warsaw Climate Change Conference, November 2013: www.unfccc.int

ARTICLES

"1997 Vintage: Proceed with Caution," L. J. Johnson-Bell, *Wine & Spirit International*, January 1998

"2012 Vintage in Emilia-Romagna," Enoteca Regionale Emilia-Romagna, September 2012

"Abruzzo: Blizzard 'Attila' Devastates Vineyards, Valentini Loses 50-Year-Old Vines," December 2012

"Alcohol Levels: The Balancing Act," Andrew Jefford, Decanter.com, March 2010

"Amplified Greenhouse Effect Shifts North's Growing Seasons," NASA, March 2013

"Australia and the Climate Criminal," Tim Flannery, *The Independent*, September 2013

"Australia: Climate Change to Threaten Vineyards," Alex Sampson, April 2013

"Australia's Wine Regions Threatened by Drought," Phil Mercer, 2008

"Bacchus Revels in a Warmer World!," Gregory V. Jones, NASA, December 2013

"Best Wines Will Come from Scotland, French Chef Says," Henry Samuel, *The Telegraph*, August 2009

"Big Story: Worst Grape Harvest in Half Century," Raf Casert, October 2012

"Boisset Wineries Go 100% Solar," *Napa Valley Register*, May 2013

"Bordeaux 2010: Alcohol Is Threatening Bordeaux Style," Panos Kakaviatos, Decanter.com, April 2011

"Bordeaux to Explore Non-Bordeaux Varieties?," Steve Heimoff, February 2009

"Burgundy Harvest 1993," L. J. Johnson-Bell, *Vintage Magazine*, December 1994

"California 1994: The Never-Ending Harvest," L. J. Johnson-Bell, *Vintage Magazine*, December 1994

"California Wine Industry Ignores Dire Climate Change Warnings," W. Blake Gray, *The Gray Report*, April 2013

"Call for Planting Restrictions to Stay Until 2030," AFP, April 2013

"Cameron 'Very Much Suspects' Climate Change Behind Recent Storms," Rowena Mason, January 2014

"Can Zero-Carbon Buildings Become a Reality?," Tim Smedley, *The Guardian*, November 2013

"Carbon Footprint of Wine Production: Climate Change Adaptation and Mitigation Possibilities in the Austrian Wine-Growing Region Traisen Valley," G. Soja et al., July 2010

"Change Is in the Air: Carbon Neutral Wine," Jacob Gaffney, May 2007

"Cheval Blanc, Clerc Milon Unveil New Green Wineries," Jane Anson, Decanter.com, June 2011

"Chile Faces Climate Change Challenge," James Painter, *BBC News*, May 2009

"Climate Change Adaptation to the Australian Wine Industry," Aysha Fleming, CSIRO, February 2013

"Climate Change and Climate Science, Q's & A's," Met Office, January 2013

"Climate Change and the Rise and Fall of Civilisations," Emily Sohn, December 13, 2013

"Climate Change and Wine: The Canary in the Coal Mine," Gregory V. Jones, January 2013

"Climate Change Impacting Worldwide Wine Production," Agence France Presse, March 2013

"Climate Change in the Vineyards: The Taste of Global Warming," Ann Cairns, Geological Society of Americas public release, November 2003

"Climate Change Is Real But We Have Time," Bjorn Lomborg, *Sunday Times*, London, September 22, 2013

"Climate Change May Cause Massive Food Disruptions," L. A. Javier and S. Li, *Bloomberg*, February 2013

"Climate Change May Cool Napa Valley, Says Study," Richard Woodward, February 2011

"Climate Change Not Responsible for Rising Alcohol Levels: Study," Hazel Macrae, June 2011

"Climate Change Rewrites World Wine List," *Discovery News*, March 2013

"Climate Change Ruining Cocktail Hour Too," Annie Bell Muzaurieta, TheDailyGreen.com, 2008

"Climate Change Threat to Australia's Top Wines," Robert Burton-Bradley, News.com.au, June 2011

"Climate Change Threatens French Wine," Marie Doezema, CNBC.com, January 2013

"Climate Change Will Produce Wine Winners and Losers," *Asian Scientist*, August 2013

"Climate Change," American Chemical Society, Public Policy Statement, 2010–2013

"Climate Change—Overview," British Antarctic Society, January 2013

"Climate Myth: It's Been Far Warmer in the Past, What's the Big Deal?,"
David Chandler, May 2007

"Coalition Turns Back on UN Climate Summit," Tom Arup, November
2013

"Current and Future Consequences of Global Change," NASA, December
2013

"Dangers of Soil Salinity," Tim Teichgraeber, *Wines & Vines*, June 2006

"Death Valley Nears Record Temperature," Tim Walker, *The Independent*,
July 1, 2013

"Dishing Out the Dirt," L. J. Johnson-Bell, *Wine Magazine*, October 2001

"Do You Think Wineries Should Focus on Adaptation Rather than
Prevention? Interview with Richard Smart and Miguel Torres,"
Gabrielle Opaz, February 2008

"Doing It His Way: Robert Skalli," L. J. Johnson-Bell, *Wine & Spirit
International*, March 1998

"Drought, Cold Slam European Vineyards," William Spain, *Market Watch*,
October 2012

"Dry, Dry Again," Larry Walker, *Wines & Vines*, November 2007

"Effect of Global Warming on Chinese Viticulture," 2007

"Effects of Deforestation on Our Climate," MetOffice.gov.uk, February
2011

"Effects of Volcanoes on Climate," MetOffice.gov.uk, February 2011

"Effects on Wineries from Global Warming and Climate Change,"
ClimateAndWeather.net, January 2013

"Ferrari Moving to Higher Vines as Climate Change Effects Felt,"
Anne Krebiehl, December 2012

"Fight to Save Wine from Extreme Weather," Jeffery T. Iverson, Time.com,
April 2011

"Fine Wine and *Terroir*: The Geoscience Perspective," R. A. Wilson,
Geoscience Canada 34, 2007

"Flooding Rains?," Karl Braganza, *The Conversation*, May 2012

"French Wine in Danger Due to Global Warming," Graham Tearse,
Decanter.com, August 2009

"Fresh Figs for Cold Climates," P. D. Forsyth, PhigBlog.com, October 2009

"Geo-thermal Heating and Cooling in Closed Greenhouse Concept," David Schmidt, July 2013

"GlassRite Wine: Bottling Wine in a Changing Climate," WRAP, May 2009

"Grape Growers Fight to Keep Planting Limits," Euractiv.com, September 2012

"Grapes, Genes and Climate Change," Africa Genome Education Institute, November 2011

"Grapes: Indoor Cultivation," Royal Horticultural Society, September 2012

"Greenhouse Gas Emission Could Be Phased Out Almost Entirely by Mid-Century," Ecofys.com, October 2013

"Greenpeace: Climate Change Could Destroy Burgundy," James Lawrence, September 2009

"Growing Apples in Florida: It Can Be Done," Jacqueline Cross, DavesGarden.com, October 2011

"Growing Grapes During the Next Grand Minimum," Russ Steele, November 2012

"Has the Garden of Eden Been Located at Last?," S. J. Hamblin, *Smithsonian Magazine*, May 1987

"Hot Temperatures Affect Wine Grape Harvest: Central Coast," Hope Hanselman, October 2012

"Hot Topic," Jane Anson, Decanter.com, August 2008

"How Do We Teach Old World Winemakers, New Tricks?," Gabriella Opaz, February 2008

"How Does 'Carbon Neutral' Fit into Organic Wine?," *Organic Wine Journal*, March 2012

"How Global Warming Could Change the Winemaking Map," Jeffery Iverson, 2009

"How Hot Is Too Hot?," Gregory V. Jones, *Wine Business Monthly*, February 2005

"How Wine Production Pollutes," News.Discovery.com, August 2011

"Ideal Irrigation Methods for Premium Wine Grapes Determined," *Science Daily*, September 2013

"In Search of Optimal Grape Maturity," *Practical Winery and Vineyard*, UC Davis, July–August 2001

"Irrigation," *Oxford Companion to Wine*

"Irrigation of Winegrapes in California," Larry Williams, *Practical Winery & Vineyard*, November–December 2001

"Is Irrigation Such a Bad Thing?," Richard Gawel

"Is the Napa Valley Heating Up? Vintners Release New Climate Study," *Napa Valley Register*, February 2011

"Is Wine Bad for the Planet?," Maggie Rosen, Decanter.com, November 2007

"Jumping to Conclusions About Climate Change," Cliff Ohmart, *Wines & Vines*, May 2007

"Kiribati Climate Change Refugee Rejected by New Zealand," *The Telegraph*, November 2013

"Land of More Extreme Droughts and Flooding Rains?" Karl Braganza, 2012

"Le SO4 ou Sélection Oppenheim 4," Institute Français de la Vigne et du Vin, 2005

"Life Cycle Emissions of Wine Imported to UK," WRAP, May 2007

"Low Yields Characterise 2012 Grape Harvest," Rosie Martin, Harpers. co.uk, September 2012

"Mesopotamian Climate Change," Megan Sever, *Geotimes*, February 2004

"Napa 'Unsuitable' for Premium Wine in 30 Years," Adam Lechmere, Decanter.com, July 2011

"Napa 2010: Long Cool Summer Promises European Vintage," Cheryl Lincoln, September 2010

Napa Valley Vintners Harvest Report 2012

"Napa Wine Quality May Suffer Climate Change," Climate Change and Wine Conference, 2011

"New Study Finds That Global Warming Could Dry Out the Southwest," Bryan Walsh, *Science & Space*, 2011

"Nyetimber Writes Off 2012 Harvest," October 2012

"Ocean Now Absorbing Heat Much Faster," Steve Connor, *The Independent*, November 2013

"Olive Oil, Sun-Dried Tomatoes and Global Warming," Rosenweig et al., NASA, March 1998

"Pinot Noir: The Quest for Cool," Stephen Brook, Decanter's California 2013 Supplement

"Plantation: Demande d'Autorisation (Formulaire)," www.inao.gouv.fr

"Preparing for the First Vintage in the Isles of Scilly," Geoffrey Kelly, Euro Strategies, April 2013

"Recommended Wine Grapes for Nova Scotia," John Lewis, AgraPoint, 2002

"Record Harvest for English Wine Makers," *The Telegraph*, January 2010

"Report: EU Planting Limits Contribute to Wine Production Slump," EurActiv, March 2013

"Rising Alcohol Levels in Wine," Jamie Goode, wineanorak.com, December 2007

"Rising Seas," Tim Folger, *National Geographic*, September 2013

"Sales of Low-Alcohol Wine Soar," James Hall, *The Telegraph*, September 2012

"Same Fruit Different Tastes," SelfSufficientMe.com, August 2011

"Seawater Greenhouse Selected as Leading Sustainable Solution," Sustainia, June 2013

"Sensational Year for English Wine," *Sunday Times*, October 2011

"South Africa Harvest Report 2013: Expecting Record Crop," VinPro, May 2013

"Spain: On the Frontline in Climate Change Battle," James Bryce, Olive Press, November 2012

"Spain's 2012 Harvest," Wines from Spain, November 2012

"Spanish Wine Makers Fight Climate Change," Danny Wood, *BBC News*, September 2008

"Supergrapes Could Make Good Wine Despite Climate Change," John Roach, *NBC News*, June 6, 2013

"Tahbilk Aspires to be World's First 'Naturally' Carbon-Neutral Winery," E. Kearnes, June 2013

"Temperature of Europe During the Holocene," *Quaternary Science Review*, 2001

"To Grow Sweeter Produce, California Farmers Turn Off the Water," Alastair Bland, 2013

"Uniform Ripening Encodes a Golden 2-like Transcription Factor Regulating Tomato Fruit Chloroplast Development," *Science*, June 2012

"Unresolved Questions About Earth's Climate," NASA, December 2012

"Updated World Map of the Köppen-Geiger Climate Classification," Hydrol Earth System Science, 2007

"USA Harvest 1995: Yields Down, Quality Up," L. J. Johnson-Bell, *Vintage*, December 1995

"Using a Hedonic Model of Solar Radiation to Assess the Economic Effect of Climate Change: The Case of the Mosel Valley Vineyards," Ashenfelter and Storchmann, National Bureau of Economic Research, 2006

"Vino e clima: la qualità si definisce nella vigna," Peppe Caridi, November 16, 2013

"Watering and Irrigation," GrowVeg.com, 2013

"What Affects Our Climate?," MetOffice.gov.uk, March 2011

"What Climate Change Means for Wine Industry," Mark Hertsgaard, April 2010

"What Does Past Climate Change Tell Us About Global Warming?" SkepticalScience.com

"What Is Climate Change?," wwww.MetOffice.co.uk, March 2012

"What Is Climate?," www.MetOffice.gov.uk, March 2012

"What Rising Temperatures May Mean for World's Wine Industry," John McQuaid, April 2013

"What's Happening to Our Climate?," Samuel Matthews, *National Geographic*, November 1976

"Why Is Our Climate Changing," MetOffice.gov.uk, January 2011

"Wine, Climate and Change," Trond Arne Undheim, *Color Magazine*, April 2013

"Winemakers Rising to Climate Challenge," Paige Donner, *New York Times*, November 2011

"Wineries Report Damage to Vines 'Minimal' in Margaret River Fires," James Lawrence, November 2011

"With Warming Climes, How Long Will a Bordeaux Be a Bordeaux?," Alastair Bland, May 2013

"You Can Fool Climate Deniers, But You Can't Fool Mother Nature: Plants Pack Up and Move North," Peter Sinclair, March 2013

INDEX